Herndon's Earth and the Dark Side of Science

J. Marvin Herndon, Ph.D.

FOREWORD by DORION SAGAN

Looked at sociologically, evolutionarily, science is a strange human enterprise. The most powerful knowledge acquisition and deployment—often in the service of technology—method the human species has ever hit upon, it shares certain human and more-than-human foibles, if not failings. Even examining it with would-be objectivity partakes of the strengths and weaknesses of science (from *scientia*, Latin for knowledge). The strengths are obvious. Traced by philosopher and mathematician Alfred North Whitehead to the ancient Greek philosophical penchant for simple, lucid speculations, combined with a more modern penchant for evidence on the ground, what William James in a letter to novelist brother Henry called "irreducible and stubborn facts,"[1] science not only boldly goes where only science and mythology have boldly gone before, but does it with a methodological commitment to change its mind if the facts so warrant. This has allowed it, unlike religion, to hit upon the correct mapping of Earth's position in space, of humankind's position in time, of the composition of our bodies and the relation of that composition to the elemental structure of the universe, and so on. Knowledge, often with mathematical precision, of electromagnetic fields, chemical reactions, and gravitational pulls has allowed human beings to commandeer vast energy resources, to voyage to the stratosphere and Moon; it has allowed our machine probes to land on Mars and to swing by the giant planets and fly completely out of our solar system; it has allowed us to understand billions of years of evolution from microbes and to understand—and weaponize—the inner workings of stars.

And yet science, dependent upon checking by scientists internationally upon the validity of one another's work, also has a democratic element of consensus—a kind of vote—that may hamper its progress. Just like an individual human being who ascribes inordinate value (often about 100% correctness) to his own opinions, the community of scientists can prematurely settle on a viewpoint that in retrospect proves not the most helpful—indeed, that in retrospect pans out to be outright wrong. This is not an occupational hazard so much as the progress and process of science itself. Science, in other words, is not perfect, but the best we have. Unlike a

---

[1] Whitehead, Alfred North, 1962. <u>Science and the Modern World,</u> Lowell Lectures 1925, The New American Library, p. 10.

religious or political opinion, or beliefs based on anecdotes that prove insufficient to make a generalization, the vetted understandings of science combine the observations and experiments of many trained professionals. But precisely for this reason, the provisional answers arrived at by science are justifiably accorded a greater truth value, a value which proves its worth, if not its absolute incontrovertible truth, by both the leads good science gives for future research and the practical applications to which it is put. Nonetheless, such merited confidence is always in danger of ossifying into orthodoxies even more powerful than the doctrines and dogmas of religion, whose knowledge one realizes is vaguer and weaker for not having been submitted to such tests. Moreover—I call this the evolutionary epistemology paradox—there is no direct correspondence between what is true and what allows an individual or society to survive— as Machiavelli and propagandists would be the first (at least in private) to argue. Thus the problem of the ossification of premature scientific knowledge can be exacerbated by the involvement of the state and private industry, both of which have their own agendas—namely, to protect and empower certain groups and help them make money—which do not always ride on the same rails as the pure scientific quest.

At the limit both these human failings of the scientific method—its tendency to overconfidence based on its own success, and its tendency to reduce a helpful diversity of theories to a single or a few accepted ones— present major problems. In the 19th century Samuel Butler, author of the iconic English novels *The Way of All Flesh* (about Victorian hypocrisy) and *Erewhon* (one of the first novels to explore the evolution of machines), gave up religion and saying his prayers on his voyage to New Zealand to become a sheep farmer. Once there, he excitedly ordered and devoured Darwin's *On the Origin of Species*. Not long after, however, he became disenchanted with what he regarded as Darwin's whitewash of earlier, phenomenologically richer accounts of evolution by Charles Darwin's predecessors, including his own grandfather, Erasmus. On the rise of doctrinaire science, Butler remarked that, "Scientists are the priests of the modern age and they must be watched very closely."[2]

---

[2] For more on the topic of scientific heretics see Chapter 11, "Priests of the Modern Age: Scientific Revolutions and the Kook-Critic Continuum, Being a Play of Crackpots, Skeptics, Conformists and the Curious, pp. 133-163 of my *Cosmic Apprentice: Dispatches from the Edge of Science*, University of Minnesota Press, 2013.

In a sense, the scientific method is deeply inimical to the "authority fever" of individual scientists because the expert, the "priest" of science, as a person with an ego, a reputation, and a job, has greater difficulty admitting he or she is wrong than the system itself. When we outsource our thinking to experts, we run the risk of sacrificing real knowledge and the search therefore to a comforting consensus that may prove deeply wrong. At the limit, the science of today may be the religion of tomorrow. And although, as the philosopher Charles Sanders Pierce has shown, there is undoubtedly an element of consensus to what we call "truth," this element—reflected in the institutionalized scientific process of "peer review," whose history is discussed in this book—can ultimately be detrimental. With regard to the second problem of science, its reduction to something everybody can agree on, Galileo writes in *The Assayer* (1623) that, "That man will be very fortunate who led by some unusual inner light, shall be able to turn from the dark and confused labyrinths within which he might have gone forever wandering with the crowd and becoming ever more entangled. Therefore, in the matter of philosophy, I consider it not very sound to judge a man's opinion by the number of his followers."

Of course I am here myself outsourcing my intellect to a kind of high priest of science, so we see the intricate human nature and philosophical difficulty presented by what might be a flawless epistemological method. Nonetheless, I think we can take home two things from this brief summary or analysis of the power and pitfalls of science. First, theoretical diversity is the lifeblood of science, and should in general be strongly encouraged, rather than put on a pedestal or presented (although not necessarily called that) as dogma. Evolution itself operates by selection among diverse variants; keeping theoretical variants open, examining them and rejecting even consensus-solidified viewpoints, is a key to long-term knowledge acquisition and deployment in ways that, again, from an inescapably human viewpoint, will enhance our species' chances of long-term happiness and ecological fitting in with extant members of the estimated thirty million other species with whom we share this planet. It thus behooves us not to foreclose entertainment of other viewpoints, sometimes even radically new and epistemologically uncomfortable viewpoints—nobody likes to admit that what he was thinking was wrong, and the longer he has been thinking it the less they he is to give it up; it has become part of his belief structure. Nonetheless, it is precisely these refreshing, new, radical—and sometimes at first crazy-seeming viewpoints—that sometimes have the power to drop, to use biblical language, the scales from our eyes.

Another thing that I think becomes clear from this brief analysis is that we need especially to be suspicious, and critical of consensus science where and when it becomes self-perpetuating by social institutions that have become compromised in their search for truth. As suggested, national and corporate agendas, which tend to valorize secrecy and eschew the aspect of international communication and openness that is crucial to the best international science, add an unhelpful layer of ossification to the vigorous search for knowledge.

At another level, but interrelated with these two pitfalls on the road to truth, are academic roadblocks erected and maintained to preserve individual egos, disciplinary boundaries, and revolving-door personal associations and funding. Ironically, as astronomer Renée James has shown in her book *Science Unshackled,* even a "pragmatic" focus on scientific research designed to solve a specific social problem (or make money) takes a backseat to pure research and curiosity. Not a straight-line need for new forensic tools but investigating pink Yellowstone microbial slime led to genetic fingerprinting; investigating twitching frog legs to the first battery; GPS systems were not found by the attempt to design them, nor WiFi; rather bother were serendipitous results, side-effects as it were, of pure research—the testing, respectively, of relativity theory using airplane-borne atomic clocks and the *failed* attempt to detect evaporating black holes.

With the following general comments behind us, we can now discuss the book at hand, the new "heretical" view of Earth as presented by J. Marvin Herndon. My mother, Lynn Margulis, a member of the National Academy of Sciences, was a fascinated advocate of Herndon's own "heresy." Her own ideas on life's symbiotic constitution and evolution were disparaged more or less until she died in 2011, at which point they became considered common knowledge and bandied about the media (usually without any mention of her name) with catchy phrases such as the microbiome. Herndon may be on a similar trajectory—on a straight-line course for posthumous recognition—at least of his ideas, if not the fact that he originated them! It is said that there are three phases in the reception of a scientific ideas: 1) You are crazy and don't know what you are talking about. 2) What you are saying is right, but it is trivial. 3) What you are saying is right, and important, but we knew it all along. Again we can see the all-too-human component of not being able to admit one was wrong in this near non-caricature of science. One might also add a fourth stage, where scientific authorities never admit that they were wrong, but eventually die—allowing for the new ideas finally be considered on their own merit.

As you will see in the following pages, Herndon is not a true outsider to geoscience. Trained in nuclear chemistry, he has published in peer-reviewed journals, been praised by Inge Lehmann, the Danish seismologist who discovered Earth's solid inner core, and studied under Harold Urey, a world expert on elemental composition of the cosmos and the thesis adviser of Stanley Miller, famous for his origins-of-life experiments at the University of Chicago. Nonetheless, Herndon's coherent, unified vision has not received the consideration it deserves. Based on available data, and a critique of the orthodox speculation of mantle convection as physically impossible, Herndon argues that Earth used to be a Jupiter-sized gas giant that lost its outer layers in T-Tauri solar wind blasts, and has since been expanding by decompression. He also argues that the Earth's magnetic field, which seems at the geological time scale to reverse polarity on the turn of a dime, is produced by natural nuclear fission in the Earth's core. These views are based on a distinct analysis of the elemental constitution and isotopic constitution of Earth's core, and are more in keeping with the observed facts of enstatite chondrite meteorite composition and the sensitivity of polar magnetic reversals. When Herndon first put forth his views there was no record of Jupiter-sized planets at one AU (astronomical unit), that is at the same distance Earth is from its star; after that discovery planetary astronomers made the rather ad hoc suggestion that such big planets could "migrate" to Earthlike distances—a proposal that violates Ockham's razor and Einstein's purported statement to make things as simple as possible but no simpler.

Herndon's ideas are also supported by another fascinating, but little-known fact, that nuclear fission can happen naturally and without any kind of human technology. This is known from fossil record of isotopes from an ancient uranium deposit in Gabon in Africa. Thus, despite consensus, insufficiently examined geoscience beliefs, like the migration of Jupiter-sized gas giants to Earth-from-sun distances, may be as unlikely as the presence of unconsidered natural nuclear fission reactors is likely. If Herndon is right, then there is no telling as yet when our geomagnetic field will fail. There is also, despite the political inconvenience of it to those who have become married to the certainty of anthropogenic global warming as an ideological litmus test, real reason to doubt the scientifico-political consensus on this issue. (This in no way means that humanity's doubling in the last fifty years is not creating serious ecological problems.)

If the georeactor fails during humanity or our technological descendant species' tenure, we will no longer have any buttress against the solar winds, and global electromagnetic telecommunications will likely fail quite quickly and dramatically. So, too, but far more aesthetically, without Earth's protective magnetic field there will be worldwide amplification of the so-called aurora borealis, colorful curtains caused by solar electromagnetic radiation but now relatively rare and confined generally to our planet's poles. Among all the apocalypses that I have heard about or scientists have entertained, Herndon's seems to be the most artistic if not poetically just.

As a final reflection let me relate that I was recently with my friend the astrobiologist David H. Grinspoon in the Library of Congress where we briefly examined (we only had fifteen minutes before closing) some of the papers of my father purchased from my father's third wife, Ann Druyan, and deposited in the Library, by Seth McFarlane, creator of the cartoon, Family Guy. There I found two objects of relevance to Herndon's quest for scientific recognition. One was a tract by Isaac Asimov in which he discussed the difficulty of recognizing new scientific ideas against the tide of a wrong consensus. At one point he spoke of scientific heretics and the virtual impossibility of the potential value of their ideas being recognized by laymen. On this (again outsourcing my intellect, as I am taking the notion from Stephen Jay Gould) I would disagree. As Gould points out, Darwin—against the great consensus of creationist orthodoxy—published his work not in a specialized journal, and not in technical language, but in a popular book, accessible to the masses, which became a bestseller. Similarly, the above-quoted Galileo, at a time when Latin was the lingua franca of science, published his book, *Dialogue Concerning the Two Chief World Systems,* in the more readily understandable common language of the people, not Latin but, to be understood by his nonspecialist fellow citizens, Italian.

The second relevant Asimov writing was more personal. He, worried (perhaps rightly, my mother said that he was arrogant and had bragged he was always first because his name appeared at the beginning of alphabetical lists) about imputations of arrogance, had corresponded with my father about the title of a new book. He was thinking of entitling his autobiography, *I, Asimov.* Needless to say, this could hardly be more of a self-tooting horn of a title. My father wrote back strongly advocating he get over his reticence to self-boosterism. Carl Sagan could think of three good reasons why to so call the book *I, Asimov*. First, and rather obviously, because it had resonance with one of Asimov's most famous science fictions, *I, Robot*. Secondly, and this I thought quite cute as well as an astute observation by my father, that when it hit the bestseller lists it would appear, "fractally" as we might now say, in *The New York Times*, as *I, Asimov* by I. Asimov. Finally—and on this point Asimov, who did not end up calling his book that, might rightfully have been more circumspect—my father seemed to indicate that he should show no compunction about naming a book about himself so directly after himself. When you think about it (I say) the title *I, Asimov* is nearly as egocentric as *Me, Myself, and I*. In any case I mention these divagations because Herndon in the following book has decided to name his popular exposition, *Herndon's Earth*...which some might consider an arrogant title. However, in his case it is arguably even more merited than in that of Asimov. In Herndon's case, unlike that of Asimov or my father, what is at issue is indeed an "heretical" view of what is now consensus science. Thus there is an argument to be made in calling attention—a cry in the wilderness if anyone will listen—to one's own potential contribution to our own understanding of our place in the solar system and universe. Herndon is going against the big guns and so far they have not even it seems taken the time to see if there are any theoretical targets to shoot at. Of all Herndon's self-published books, this one most dramatically tells them what to look for, and why. Herndon deserves to have his new picture of Earth and our solar system considered carefully and responded to meticulously and publicly by the experts in the field. In the meantime it takes his case—for the sake of unadulterated scientific curiosity, unbeholden to institutional interests—directly to the intelligent public, that they may get a taste of what may, at some future date, be considered accepted knowledge. © Dorion Sagan, 2014

PREFACE

Now, as in ancient times, we share the common quest to understand our world and its place and origin in the cosmos. Although we live in a technologically advanced society, the origin and nature of our own planet has to a large extent been a great mystery. On the one hand, we live for just a brief moment in time on an ancient, complex, and largely inaccessible planet. On the other hand, scientists are increasingly becoming more narrowly focused specialists. Beneficial as this may be in terms of honing expertise via various disciplinary inquires, it nevertheless limits our ability to understand the ever-important big picture. As a consequence of this specializing tendency, fundamental advances, paradigm shifts, despite their importance, tend to come about slowly and infrequently. They have however been made, in the geosciences for example by Vladimir Vernadsky (1863-1945) in Russia and Eduard Suess (1831-1914) in Austria, and will continue to be made albeit not very often.

Many consider the continental drift theory of Alfred Wegener (1880-1930) to be the paradigm shift that eventually matured to plate tectonics theory. In setting forth that theory, Wegener used the title *The Origin of Continents and Oceans* [1, 2]. Why, you might ask, did he not use the title *Wegener's Earth*? Perhaps, one might guess, it is that others before him since the time of Abraham Ortelius (1527-1598) had proposed continent displacement for a variety of reasons or perhaps because Wegener's theory was incomplete, lacking a plausible mechanism and an adequate energy source. Possibly. But I doubt Wegener would have considered such a title as he was fully aware of the limitations of continental drift theory, and I might add plate tectonics theory, as indicated by his statement, page 167 [2]:

> **The determination and proof of relative continental displacements, as shown by the previous chapters, have proceeded purely empirically, that is, by means of the totality of geodetic, geophysical, geological, biological and paleoclimatic data, but without making any assumptions about the origin of these processes. This is the inductive method, one which the natural sciences are forced to employ in the vast**

majority of cases. The formulation of the laws of falling bodies and of the planetary orbits was first determined purely inductively, by observation; only then did Newton appear and show how to derive these laws deductively from the one formula of universal gravitation. This is the normal scientific procedure, repeated time and again. The Newton of drift theory has not yet appeared.

Now, you might ask, why *Herndon's Earth*? Humbly, I would say, this is the reason: For the first time ever, from Earth's early formation as a Jupiter-like gas giant, I am able to derive virtually all of the geological and geodynamic behavior of our planet, including the origin of mountains characterized by folding, the primary initiation of fjords and submarine canyons, the origin of continents and oceans, seafloor topography without mantle convection, Earth's internal composition, its previously unanticipated and potentially variable energy sources, its variable magnetic field, petroleum and natural gas deposits, and the non-anthropogenic basis of atmospheric carbon dioxide increases.

There is also another reason for the title *Herndon's Earth*. In 1993, I demonstrated the feasibility of a natural nuclear fission reactor at Earth's center, called the georeactor, one of the consequences of Earth's early formation as a Jupiter-like gas giant. With confirmation of my calculations and with subsequent developments, other georeactors at places within the Earth other than its center were suggested and published. However, none of these "copy-cat" georeactors were potentially viable as, lacking confinement, they would melt down to Earth's center, the location of, now probably better to say, "Herndon's georeactor," to distinguish it from these other impracticable reactors.

But what about the rest of the title, the *Dark Side of Science*? I didn't wake up one morning with the idea of Earth having originated as a gas giant. In the 1970s I began to develop the necessary understanding, for example, by realizing that the inner core of the Earth, an object about the size of the Moon and three times as massive, might have a composition unlike the idea that had prevailed for 40 years. In 1979, I published the new idea in the *Proceedings of the Royal Society of London* [3], but instead of engendering debate and discussion, it was ignored and its proposer was arguably "excommunicated." In science one often makes discoveries, especially about the big picture that have many downstream consequences for individual studies by asking the question: "What's wrong with this picture?" With respect to scientists' behavior, I asked the same question and I share here what I have learned about the forces that have, for decades been diminishing American science capability, corrupting individuals and the institutions they serve, and that have led geoscientists into becoming part of what I have named "the malevolent political agenda." Now it should be said not all the participants in a malevolent political agenda are themselves consciously malevolent; they may be lazy, ignorant, or compromised by the need to protect their careers, funding, or social milieu. Nonetheless, we must recognize that the political agenda in which they are involved is itself malevolent, taking on a life of its own, to the detriment of the methods of science and the search for truth, not to mention the national progress and public safety.

Scientific specialization can be like an edifice built on a corrupt foundation or the proverbial house of cards; standing seemingly sturdily or precariously for a while but becoming increasingly unviable with each new wing or addition. This has been the case, in my not-so-humble judgment, with the growth of geoscience. Geoscientists have steadfastly been building upon a half-century old misunderstanding of Earth's behavior which is considerably different from the behavior that is a consequence of Earth's early formation as a Jupiter-like gas giant. The ramifications are all about us, but none are as ubiquitous as the idea that increasing amounts of carbon dioxide in the atmosphere are anthropogenic (human produced) and are causing global warming via the greenhouse effect. Rather, I would argue that variability of Earth's major energy sources, not previously considered, is responsible for the observed increase over at least the last 40 years in the annual number of earthquakes of magnitude ≥ 6 and for changes in seawater temperature which is directly related to atmospheric carbon dioxide composition. The relationship between seawater temperature and atmospheric carbon dioxide is evident over a period of 800,000 years from measurements of an Antarctic ice core:

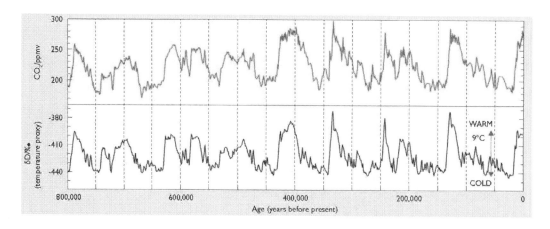

Correlation between local temperature and trapped carbon dioxide (CO_2) over a time span of 800,000 years from an Antarctic ice core (see Figure 1.5). Reproduced with permission of the British Antarctic Survey.

One may now say, with regard to mainstream, increasingly strident, ideological, and politicized comments on anthropogenic (human caused) global warming, variability of atmospheric carbon dioxide content is related to variable Earth-heat production, a consequence of Earth's early formation as a Jupiter-like gas giant. Human activity, I submit, is not to blame.

This all-too-brief introduction does not begin to reveal the connections between Earth's origin and the basis of similarities and differences of other planets in our Solar System, or how stars including our own ignite, or the reason why the multitude of galaxies display just a few prominent patterns of luminous stars. So, please enjoy *Herndon's Earth and the Dark Side of Science.*

Preface References

1.   Wegener, A., *Die Entstehung der Kontinente und Ozeane.* fourth ed1929, Braunschweig: Friedr. Vieweg & Sohn. 246.

2.   Wegener, A., *The Origin of Continents and Oceans.* Translated by John Biram1966, New York: Dover Publications, Inc.

3.   Herndon, J.M., *The nickel silicide inner core of the Earth.* Proc. R. Soc. Lond, 1979. A368: p. 495-500.

TABLE of CONTENTS

1. Science Hijacked: A Malevolent Political Agenda

This truth is incontrovertible. Panic may resent it, ignorance may deride it, malice may distort it, but there it is. — *Colonel Winston Churchill, Royal Assent, Hansard, May 17, 1916*

Science, from the Latin *scientia*, meaning knowledge, is meant to be an objective practice, however, it necessarily, as a group, cultural enterprise has subjective elements. These apply not only to its methodology, as interpretations of quantum mechanics underscore, but also, and more to the point of the present monograph, to its corruptibility by political and corporate interests. Science, whose knowledge acquisition comes largely from its ability to self-correct, in essence admit that it's wrong in the face of the evidence, has shown itself to have immense power, as even a cursory glance at the history of WWII, decided by the development and deployment of nuclear bombs will show. I write the following as a nuclear geophysicist who has published multiple peer-reviewed papers in world-class journals, who studied with Harold Urey (whose student Stanley Miller performed the first origins-of-life experiments), and whose advocacy for a more coherent view of Earth science based on more comprehensive evidence was praised by Inge Lehmann, the Danish seismologist who discovered the Earth's inner core in 1936. As Thomas Kuhn showed in his well-received *The Structure of Scientific Revolutions* (and before him, Ludwig Fleck in his *The Genesis and Development of a Scientific Fact*), radical revisions of scientific theory do not happen easily, or on time. They await a critical mass to be realized, as human nature remains stubborn and it is easier to keep on believing what one has believed — and the longer it has been believed the more the ego is invested in it — than to change one's mind in the face of evidence. These problems, already profound in the reception of new scientific ideas, only become exacerbated in our age of political control and scientific bureaucracy, where keeping one's job, or fitting in, may be more powerful criteria for the working scientist than his or her true calling, the search for truth. I have been called a maverick, and a kook. The purpose of this book is to cut through such labels and take my work — which is severely critical of what Orwell called "group-think" — on *both* sides of the political spectrum—

directly to the reader. That reader is you. My goal thus is twofold: to show the implications of my heterodox (but of course, I believe, true) ideas on geoscience; and to show how they as well as science in general (and of course, me personally) have been afflicted by a global political climate increasingly — and dangerously — hostile to the search for truth which is science's only *raison d'être*.

We may, with only slight exaggeration or idealization, distinguish between good politics and bad politics, the former of which provides fertile ground for science. What then should be the agenda of a *benevolent* political party? I say it should be essentially no different than the expressed intent of the *Preamble to the United States Constitution*.

> **We the people of the United States, in order to form a more perfect union, establish justice, insure domestic tranquility, provide for the common defense, promote the general welfare, and secure the blessings of liberty to ourselves and our posterity, do ordain and establish this Constitution for the United States of America. –** *Preamble to the United States Constitution*

The agenda of a *malevolent* political party, on the other hand, is characterized by far less altruistic ideals that may include financial and physical harm to citizenry, injustice, treachery, deceit, and worse. There are numerous instances in history when a malevolent political party takes power by force, and then perpetuates its particular penchant for Evil through the instrumentality of a police state. But what of the instance when a democratic republic is conquered from within by a malevolent political party that becomes just as depraved and brutal? Philosophers, historians, revolutionaries, and others have considered the subject from broad and detailed perspectives. My more narrowly focused interest in the latter instance is this: How does the agenda of a malevolent political party differ from the agenda of a benevolent party and, in that connection, what is the role of the scientist and the university?

In my view, a malevolent political agenda is one that embraces an intellectually-warped intent that lies outside of the scope of good governance and beyond the realm of scientific integrity. Lies and deception abound. Endorsement by university scientists provides the deceptive aura of credibility, respectability, and responsibility. Warped extra-governance intent may provide a well-spring of mal-opportunities. For example, it may serve as a basis to: broaden appeal to constituencies; consolidate political power; enrich special interests; exert control over industries and individuals, expand police powers; redistribute wealth; propagandize the young; enslave and/or annihilate citizenry; and more. Inevitably, the ultimate consequences are devastating to humanity.

There is perhaps no better example of the willing involvement by scientists in a malevolent political agenda than in the warped political agenda of Nazism (National Socialism; in German, Nationalsozialismus). Unspeakable atrocities were committed in the name of improving the German race and eliminating those who were considered "unfit." Nazi actions, which included, but were not limited to, sterilization, euthanasia, perverted medical experiments, racially-based mass murder, murder of children and the handicapped, sanctioned brutality, confiscation of property, dismissal from employment, and blacklisting of businesses, had been given a "scientific" stamp of approval by obliging, but intellectually corrupt German university professors and administrators.

There is strong evidence that scientists actively designed and administered central aspects of the Nazi racial policy [1]. In 1921, Erwin Baur, Eugen Fischer, and Fritz Lenz, university professors and/or institute scientists, published the two-volume work, *Grundriß der menschlichen Erblehre und Rassenhygiene* (*Outline of Human Genetics and Racial Hygiene*) [2], which codified the then popular Darwinian misunderstanding. The publisher gave a copy of the 1923 second edition to Adolph Hitler in prison, who read it and incorporated its ideas into *Mein Kampf*. Later, authors of official commentaries on Nazi racial laws quoted that work as their scientific basis. In 1923, the University of Munich established the first German chair in race hygiene. By 1932 there were more than 40 courses on race hygiene being offered at German universities. When the Nazis came to power on January 30, 1933, racial hygiene chairs were established at nearly every German University. During the 1930s, the Nazi regime issued numerous laws and regulations to implement its eugenic racial program; the practitioners of race hygiene, the anthropologists, geneticists, psychiatrists, and physicians, were involved in drafting and applying those laws and regulations. Before Hitler came to power, only about thirteen full professors had joined the Nazi Party. But soon after Hitler became Chancellor, hundreds showed their support in a flood of publications, speeches, and correspondence that included praise for Nazi Party policies [3]. For detailed documentation of scientists' active involvement in the Nazi malevolent agenda, from which some of these examples are abstracted, I recommend the book, *The Origins of Nazi Genocide From Euthanasia to the Final Solution*, by Henry Friedlander [4].

The present day, now-ongoing "scientifically sanctioned" malevolent agenda, referred to as "anthropogenic (human caused) global warming", is based upon science that is just as flawed as that of race hygiene. To many readers this may seem like an outrageous, ideologically motivated statement. If they are left-leaning this may especially be the case, and they may be tempted to fling this book across the room in disgust at the ravings of a presumed crank. If on the right wing (but not too far) they may nod in self-satisfaction that a PhD apparently shares their opinion. I discourage both these knee-jerk responses, value judgments that themselves discourage further investigation. As scientists, we want to focus on the evidence, and be willing to change our minds on them whatever our political stripe. We must have not the courage of our convictions so much as the courage to go against our convictions, to paraphrase the philosopher Nietzsche. Moreover, as science becomes increasingly politicized and corporatized, we must be willing to objectively investigate cultural currents that impede the free flow of information which is the scientist's lifeblood.

Sometimes scientists begin to explain observations, not realizing or admitting that there may be a different, more fundamental explanation that is not apparent to them, as I explain in this book. Just as "racial purity" was the flaw of the Nazi agenda, "anthropogenic" is the flaw in the global warming agenda: There is, and I describe here, a more fundamental explanation for the correlation between temperature and atmospheric carbon dioxide over the past 800,000 years that does not relate to human activity.

The "global warming" agenda had its beginnings in the 1980's, especially with the 1988 formation of the Intergovernmental Panel on Climate Change (IPCC) by the United Nations. The first report by the IPCC in 1990 claimed that the world has been warming and that future warming seems likely; the supposed culprit being anthropogenic, human caused, additions to the atmosphere of carbon dioxide (CO_2), allegedly causing a "greenhouse" effect. Then, along came the modelers (many of them basing their models on other models), with grand climate models based upon the false assumptions that heat from the Sun and heat from within the Earth are both constant. With those predominant variables unrealistically held constant, the tiny greenhouse effect allegedly resulting from anthropogenic increases in atmospheric carbon dioxide might appear significant. The intended result of those climate models is to demonstrate that human activities are indeed causing global warming and that the consequences are dire, threating our entire planet and its very life-forms. Now with this background a malevolent political party, (according to my definition above and not, of course, necessarily realizing they fit that definition; indeed, they think they are doing good) will save humankind and the planet: That is, I would argue, one of the malevolent political agendas of our time, and it — despite the difference in political "wings" and even initial apparent absurdity of the claim, a favorite of over-the-top rhetoricians who want to pillory targets by comparing them to a universal bogeyman — indeed shares elements in common with the Nazi malevolent political agenda.

- **For the Nazis, the "enemy" was the "unfit" who were allegedly responsible for Germany's problems and weakening the Aryan gene pool: Now, it is the burners of fossil fuels, especially coal-burning electricity-producers, who are polluting the atmosphere with carbon dioxide (CO_2) and allegedly causing global warming by the greenhouse effect.**
- **For the Nazis, laws and regulations were established to eliminate those considered "unfit": Now, one of the gases we exhale, carbon dioxide (CO_2), is deemed a pollutant, and laws and regulations limiting its production are being enacted.**
- **For the Nazis, the "scientific justification" of race hygiene was badly flawed: Now, the "scientific justification" of anthropogenic global warming is badly flawed.**
- **For the Nazis, fear mongering propaganda such as by Joseph Goebbels and his acolytes, *e.g.*, the movie, *Der ewige Jude*, was a tool: Now, fear**

mongering propaganda by Al Gore and his acolytes, *e.g.*, the movie, *An Inconvenient Truth*, has been deployed.

- For the Nazis, indoctrination of school children with racial propaganda including the movie *Der ewige Jude*: Now, indoctrination of school children with global warming propaganda including the movie *An Inconvenient Truth*. One may certainly object that the Nazi propagandists were more cognizant of their own use of scientific propaganda, and that Gore *et al.* actually think they are on a cultural mission to genuinely "save the planet"; however in both cases, as I will argue, scientific evidence and analysis has been perverted, deeply, by ideological agendas.

The Climate Research Unit at the University of East Anglia in the United Kingdom (CRU) was at the forefront of global warming climate models. Then, beginning on November 19, 2009, a whistle-blower leaked thousands of emails and documents from CRU [5] that appeared to validate earlier allegations [6] of science suppression, misrepresentation, and data-altering. In response to the scandal, called "Climategate", almost everywhere the phrase "global warming" was replaced with "climate change". With a new name, the global warming agenda nevertheless continued and was extended to include "geoengineering" the entire Earth to compensate for supposed human-caused global warming, and also to include specific high-tech weather modification activities [7-10].

About Climategate [5], as noted in a November 24, 2009 *Wall Street Journal* article:

> Yet even a partial review of the emails is highly illuminating. In them, scientists appear to urge each other to present a "unified" view on the theory of man-made climate change while discussing the importance of the "common cause"; to advise each other on how to smooth over data so as not to compromise the favored hypothesis; to discuss ways to keep opposing views out of leading journals; and

to give tips on how to "hide the decline" of temperature in certain inconvenient data.... When deleting, doctoring or withholding information didn't work, Mr. Jones suggested an alternative in an August 2008 email to Gavin Schmidt of NASA's Goddard Institute for Space Studies [GISS], copied to Mr. Mann. "The FOI [Freedom of Information] line we're all using is this," he wrote. "IPCC is exempt from any countries FOI—the skeptics have been told this. Even though we . . . possibly hold relevant info the IPCC is not part of our remit (mission statement, aims etc) therefore we don't have an obligation to pass it on"....we do now have hundreds of emails that give every appearance of testifying to concerted and coordinated efforts by leading climatologists to fit the data to their conclusions while attempting to silence and discredit their critics.

So, what was NASA's role in the global warming deception? And what should NASA's role have been? Science is about telling the truth, the full truth. With its global temperature sensing satellites, NASA should have maintained the highest standard of scientific integrity and objectivity. Instead, NASA did the opposite and became part of the malevolent agenda. NASA became a willing contributor to the global warming deception. There is no surprise here: Such behavior is entirely in keeping with NASA's politically-based organizational culture that does not follow scientific standards.

Since 1979, NASA has been using satellites to measure air temperature of the lower troposphere directly above the Earth's surface. The global satellite temperature data has been validated by radiosonde (radio-telemetry) weather balloon measurements. Instead of just relying upon thirty years of validated global coverage satellite data, the gatekeeper of NASA's climate data chose instead to use the problematic networks of land and ocean based sensors.

The NASA Goddard Institute for Space Studies (GISS), a component laboratory of NASA's Goddard Space Flight Center, conducts research that *"emphasizes a broad study of Global Change, the natural and anthropogenic changes in our environment that affect the habitability of our planet"* as its home page states. From 1981 to 2013, GISS was directed by James E. Hansen, who essentially served as the *de facto* gatekeeper for NASA's climate related models and underlying data.

In science laboratory courses, students generally are taught the importance of objectivity and the necessity to eliminate bias. So, one might reasonably expect a NASA laboratory to be directed by an unbiased individual. But, that was not the case for NASA's GISS: James Hansen has a long and very public record of being a pro-global warming activist.

From an article in the *Wall Street Journal* Aug. 29, 2007:

> **What's more disturbing is what this incident tells us about the scientific double standard in the global warming debate. If this kind of error were made by climatologists who dare to challenge climate-change orthodoxy, the media and environmentalists would accuse them of manipulating data to distort scientific truth. NASA's blunder only became a news story after Internet bloggers played whistleblower by circulating the new data across the Web.**
>
> **So far this year NASA has issued at least five press releases that could be described as alarming on the pace of climate change. But the correction of its overestimate of global warming was merely posted on the agency's Web site. James Hansen, NASA's ubiquitous climate scientist and a man who has charged that the Bush Administration is censoring him on global warming, has been unapologetic about NASA's screw up. He claims that global warming skeptics – "court jesters", he calls them -- are exploiting this incident to "confuse the public about the status of knowledge of global climate change, thus delaying effective action to mitigate climate change."**
>
> **So let's get this straight: Mr. Hansen's agency [NASA] makes a mistake in a way that exaggerates the extent of warming, and this is all part of a conspiracy by "skeptics"?**

As noted in a January 1, 2009 *Heartlander Magazine* article by James M. Taylor:

> **Many climate scientists have criticized GISS in recent years for routinely claiming significantly higher global temperatures than those reported by other scientists; for employing a staff that appears to see its role more as advocates than as scientists; for getting caught claiming recent years were warmer than the data indicated; and for failing to provide transparency in how they manipulate raw temperature data before presenting their adjusted "official" temperature reports.**
>
> **After GISS generated substantial media attention with its claim October 2008 was the warmest October in history, a number of global warming "skeptics" smelled something fishy and examined the data themselves. They soon discovered NASA and its partners at the National Oceanic and Atmospheric Administration had copied the September 2008 temperature data from Russia into the October Russian temperature dataset.**

These examples reveal part of a broader pattern of NASA's questionable "global warming" reporting that "skeptics" have revealed. So, one might ask, why did NASA management not re-assign Hansen to a position where his bias would not appear to influence or distort NASA's science? Why? Perhaps because Hansen's activities helped to insure that NASA had the E-ticket ride on the global warming gravy train, that NASA was an active participant in a malevolent agenda which affords the malevolent political party grand opportunities to exert political control over vast segments of America's industries and citizens, to enrich party insiders at taxpayer expense, and more.

The very flawed model-based assertions of anthropogenic (human caused) global warming, as well as the alleged data-tampering and fear-mongering by Al Gore, NASA, and others has led many non-scientists to consider "global warming critics" as anti-environmentalists. But, that is a gross misperception. Instead, the current extension of the anthropogenic global warming agenda to geoengineering arguably poses one of the greatest potential threats to the environment ever caused by humans, potentially poisoning biota, causing widespread crop failures, and worse. Environmentalists, like so many others, have been duped into supporting a malevolent political agenda that, as I describe in this chapter, may ultimately wreak destruction, not only to the very environment they would protect, but to the wellbeing of humanity as well.

How is an ordinary person, not a trained scientist, to make a judgment on the legitimacy of global warming, especially in light of the scientific controversy? Let me explain. In science there should be truthful debate, discussion, and controversy; an open airing of diverse ideas and interpretations generally leads to better understanding and to the forward progress of science. Agenda-driven scientists, on the other hand, identity themselves by their willingness, individually or with others, to: lie, deceive, besmirch their opposition, engage in science suppression, misrepresent, distort and/or falsify data. Moreover, by these practices they identify the agenda as being a malevolent agenda perpetrated by a malevolent political party.

Who are those besmirched as "skeptics", "deniers", and "court jesters?" Through tedious efforts they have uncovered the deceit, misrepresentation, and "erroneous" data used to advance the malevolent agenda. The literature on the subject is voluminous, so much so, that it is easy to lose sight of the forest for the trees. My approach is entirely different: I begin at the foundations of knowledge, seek to understand the commonalities, and then proceed to discover the underlying mistakes that have been made. Mistakes, fundamental mistakes, have indeed been made.

In 1957, Roger Revelle and Hans E. Suess published a scientific paper, entitled "Carbon dioxide exchange between atmosphere and ocean and the question of an increase of atmospheric CO_2 during the past decades" [11]. In that paper Revelle and Suess suggest that the Earth's oceans would absorb excess carbon dioxide generated by humanity at a much slower rate than previously predicted by geoscientists and might create a "greenhouse" effect that might eventually lead to global warming.

Twenty years after publication of the Revelle-Suess paper, over a period of four years, I had lengthy discussions with Suess about the possibility of global warming and the necessity to validate the effect of increased atmospheric carbon dioxide in a broader framework that includes the potential variability Earth's major heat source, the Sun, the factors potentially affecting the variability of Earth's reflectivity and its consequences, the potential variability of heat being brought to the surface from deep within the Earth and its potential impact on ocean warming and on changes in ocean circulation.

Since 1939, scientists have been measuring the heat flowing out of continental-rock [12, 13] and, since 1952, heat flowing out of ocean floor basalt [14]. Continental-rock contains much more of the long-lived radioactive elements than does ocean floor basalt. So, when the first heat flow measurements were reported on continental-rock, the heat was naturally assumed to arise from radioactive decay. But later, ocean floor heat flow measurements, determined far from mid-ocean ridges [15], showed more heat flowing out of the ocean floor basalt than out of continental-rock measured away from volcanic heat-producing areas [16]. For decades, heat flowing out of the Earth has been assumed to result from the natural radioactive decay of ^{235}U, ^{238}U, ^{232}Th, and ^{40}K and from heat left over from (assumed) planetesimal Earth formation [17]. Because of the extremely long half-lives involved, on a human time-scale the heat flow from within the Earth was thought to be constant. This constancy of heat flow is a fundamental assumption of all climate models, but it is extremely unlikely to be correct.

For new readers the following argument may not make the specific arguments about politicized science I am making any easier to accept. Indeed, it may only add to any group-think tendency to dismiss both the received views of geoscience and the connected, received views of political action required to combat supposed anthropogenic global warming/climate change. However, my critique of politicized science is integrally connected to my critique of mainstream geoscience, a more comprehensive alternative to which is at the core of this book.

I have disclosed a new indivisible geoscience paradigm [18-20], called Whole-Earth Decompression Dynamics (WEDD), that begins with and is the consequence of our planet's early formation as a Jupiter-like gas giant and which permits deduction of:

- **Earth's internal composition and highly-reduced oxidation state [21-23];**
- **Core formation without whole-planet melting [24];**

- **Powerful new internal energy sources, protoplanetary energy of compression and georeactor nuclear fission energy [25];**
- **Mechanism for heat emplacement at the base of the crust [26];**
- **Georeactor geomagnetic field generation [25, 27];**
- **Decompression-driven geodynamics that accounts for the myriad of observations attributed to plate tectonics without requiring physically-impossible mantle convection [28, 29];**
- **A mechanism for fold-mountain formation that does not necessarily require plate collision [30], and;**
- **A mechanism for the formation of fjords and submarine canyons [30].**

Of interest here in connection to what I am calling the "malevolent agenda" are the new concepts for heat production and its transport.

In bold contrast to the Earth-heat constancy assumption underlying climate models, Terracentric georeactor nuclear fission produces potentially variable heat that is channeled to the surface [31]. Moreover, a portion of the stored energy of protoplanetary compression, emplaced as heat at the base of the crust by mantle decompression thermal tsunami [26], is likewise potentially variable. These two variable heat sources link directly to ocean water.

The geoscience community has been locked into Earth formation/dynamics models that originated in the 1960s and 1970s and which are based upon the assumption of constant Earth-heat production [32, 33]. Consequently, there have been no systematic investigations that seek evidence and understanding of variable Earth-heat production, but there should be.

Thermal structures, sometimes called mantle plumes, lie beneath the volcanic islands of Hawaii and Iceland, whose lava contains traces of helium that appear to harbor the signature of georeactor-produced heat and helium, *i.e.*, high $^3He/^4He$ [34]. The heat source for these two stable volcanic "hotspots," as imaged by seismic tomography [35, 36], extend to the interface of the core and lower mantle, further reinforcing their georeactor-heat origin. In 2009, Mjelde and Faleide [37] discovered a periodicity and synchronicity through the Cenozoic in lava outpourings from Iceland and the Hawaiian Islands, hotspots on opposite sides of the globe, that Mjelde *et al.* [38] suggest may arise from variable georeactor heat-production.

Despite orthodox views, virtually all geodynamic activity and surface geomorphology appear more coherently explained as the consequence of whole-Earth decompression augmented by georeactor produced heat [30, 39]. Earthquakes are a consequence as well, which helps us to understand why earthquakes sometimes occur at depths as great as 660 km and sometimes occur in the middle of continents.

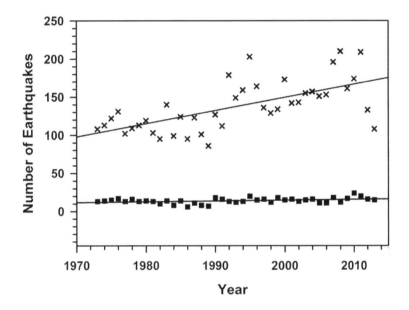

Figure 1.1 The annual number of global earthquakes, magnitudes ≥ 6 and ≥ 7, from the U. S. Geological Survey database, shown with linear regression fit lines. For the ≥ 6 magnitude data set, the coefficient of time (t) (1.727) is the slope of the line. It is statistically significant, t=5.121, p value < .001. This means that the number of earthquakes of magnitude ≥ 6 increases by 1.727 each year. This figure clearly shows that there has been a dramatic increase over the time in the interval 1973-2013 for earthquakes of magnitude ≥ 6.

Shearer and Stark [40] utilized an elaborate data selection mechanism to show that earthquakes of magnitude ≥ 7 have not increased in recent years. In Figure 1.1, I essentially confirm their result using unselected annual totals in the same magnitude range, albeit with a paucity of data. Further, Shearer and Stark, reasoning from within a flawed geoscience paradigm, remark: "Moreover, no plausible physical mechanism predicts real changes in the underlying global rate of large events." In the new indivisible geoscience paradigm, called Whole-Earth Decompression Dynamics (WEDD), variability in the Earth's two principal internal energy sources leads to considerable geophysical changeability, including potential earthquake number and/or intensity variability. As I show in Figure 1.1, there has been a dramatic increase in the number of earthquakes of magnitude ≥6 over the interval 1973-2013. Actually the increase is observed since 1900, when earthquake data became available, but I consider the 1973-2013 range to be less-error prone and more robust than earlier years due to the extensive seismic networks operant during that period for monitoring nuclear explosions.

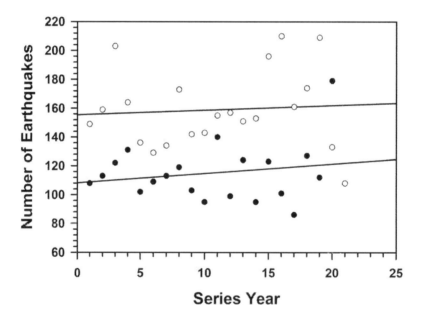

Figure 1.2 The annual number of global earthquakes of magnitude ≥ 6, from Figure 1, divided into two time intervals, 1973-1992 and 1993-2013, with the first year of each designated year 1. The 1993-2013 data points clearly lie above those of the earlier 1973-1992 data set, and both regression-line slopes are positive. This data presentation reinforces the conclusion from Figure 1.1, namely, a progressive increase in the number of magnitude ≥6 earthquakes over time.

For internal comparison, I split the magnitude ≥6 earthquake data into two sets with ranges 1973-1992 and 1993-2013, relabeling the first year of each set as year 1; the plots are shown in Figure 1.2. The calculated means for the two data sets are different. The 1993-2013 data points clearly lie above those of the earlier 1973-1992 data set, and both regression-line slopes are positive. This indicates a progressive increase in the number of magnitude ≥6 earthquakes, unless there is some unknown systematic error in the United States Geological Survey data base.

Instead of assuming constant Earth-heat production as has been done for more than half a century, one should consider and investigate Earth-heat variability. The fundamental implication of Earth-heat variability is ocean temperature variability which directly affects atmospheric CO_2 variability.

The ocean is the major reservoir for carbon dioxide, CO_2. Carbonate is a weak acid-base system existing in the ocean as dissolved carbon dioxide, carbonic acid, bicarbonate ions and their complexes [41]. In seawater, dissolved carbon dioxide, $[CO_2]$, neglecting minor forms, is:

$$[CO_2] = [CO_2(aq) + H_2CO_3]$$

In thermodynamic equilibrium, atmospheric (gaseous) carbon dioxide, $CO_2(g)$, and seawater $[CO_2]$ are related by Henry's law:

$$CO_2(g) \overset{K_0}{=} [CO_2]$$

where K_0, the solubility coefficient, is a function of temperature and salinity. Using an equation for K_0 derived by Weiss [42], I calculated the values of the CO_2 solubility coefficient, K_0, throughout the entire range of temperatures and salinities relevant to seawater (Figure 1.3). From this figure one thing is clear: An increase in temperature, over all ocean conditions, leads to a decrease in CO_2 solubility and, hence, an increase in atmospheric CO_2.

In ordinary language, the above paragraph says this:

Most of Earth's carbon dioxide resides in seawater, and there is an inherent relationship between the amount of carbon dioxide in the atmosphere and the temperature of seawater.

- An increase in seawater temperature, leads to an increase in atmospheric carbon dioxide and a decrease in seawater carbon dioxide.
- A decrease in seawater temperature leads to a decrease in atmospheric carbon dioxide and an increase in seawater carbon dioxide.

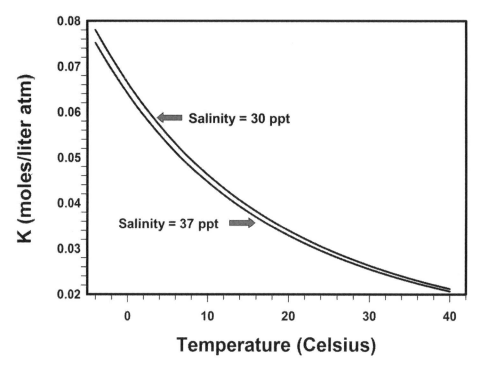

Figure 1.3 The CO_2 solubility coefficient, K_0, calculated throughout the entire range of temperatures and salinities, bracketed by the two extreme salinities, relevant to seawater. This figure shows that an increase in temperature, over all ocean conditions, leads to a decrease in CO_2 solubility.

I have made calculations that illustrate the amount of CO_2 change that can result from a small percent increase in seawater temperature, from T to T+δT. As seawater $[CO_2]$ is the primary reservoir for CO_2, I assumed for these calculations that $[CO_2]$ remains unchanged essentially unchanged by a small change in temperature [43].

The calculations were made for a series of seawater temperatures from 5°C to 40°C. The results, presented in Figure 1.4, show that just a modest increase in seawater temperature is sufficient to cause a significant increase in atmospheric CO_2. These were not intended as model calculations; the oceans are far too complex for that undertaking. Rather, they were undertaken to demonstrate the feasibility of my assertion that Earth-heat variability leads to seawater temperature variability which is mainly responsible for atmospheric carbon dioxide variability.

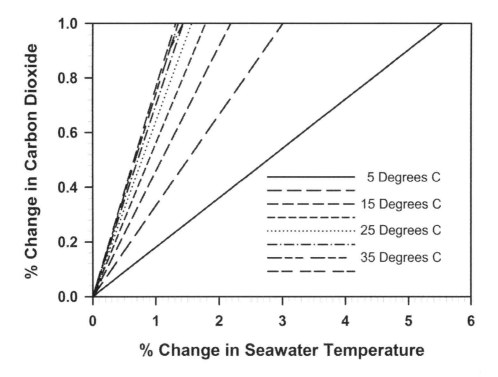

Figure 1.4 Results of calculations [43] for a series of seawater temperatures from 5°C to 40°C, which show that just a modest increase in seawater temperature is sufficient to cause a pronounced increase in atmospheric CO_2.

Antarctica boasts the longest continuous ice core, drilled at Dome C on the Antarctic plateau. The core contains a record of the air that was trapped within the ice and other information, including the temperature in the environment at the time of deposition, which can be determined from the isotopic composition of the ice. Figure 1.5 shows the CO_2 content of trapped air bubbles in the core, going back 800,000 years, and a proxy for deposition temperature, indicated by the amount of deuterium, ^2H, in the ice core relative to seawater. Despite uncertainties as to whether ice-core temperature and CO_2 are precisely contemporaneous, nevertheless, the environment temperature from deuterium correlates quite well with atmospheric CO_2 levels over the entire 800,000 years represented by the core.

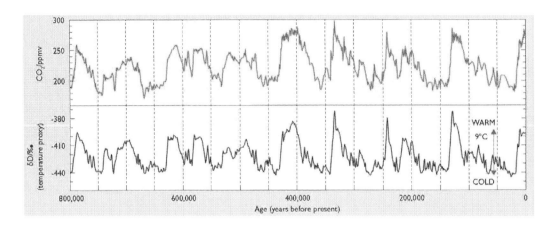

Figure 1.5 Correlation between local temperature and trapped CO_2 over a time span of 800,000 years from an Antarctic ice core [53]: Ice core data from the EPICA Dome C (Antarctica) ice core: deuterium is a proxy for local temperature; CO_2 is from the ice core air. Data from [54, 55]. Reproduced with permission of the British Antarctic Survey.

The correlation shown in Figure 1.5 suggests a causal relationship:

> **Variable Earth-heat production caused by georeactor variability and decompression-generated heat variability, causes variable ocean temperature which causes variable seawater CO_2 solubility and, hence, variable atmospheric CO_2. Variable Earth-heat production also causes variable atmospheric temperature and can lead to changes in ocean water circulation. Said another way, climate change is to a great extent caused by changes in Earth-heat production, not by the anthropogenic CO_2 greenhouse effect.**

Wrongly assuming constant Earth-heat production, on the other hand, falsely implies the temperature variation, such as expressed in this figure, occurs as a consequence of the atmospheric greenhouse effect, and begs the question as to the fundamental cause of CO_2 variability, especially over 800,000 years when human CO_2 "pollution" has only occurred in a major way for less than 150 years.

All global warming models, now called climate change models, are based upon the assumption that annually and globally seawater temperature does not change because of the underlying assumption that Earth-heat is constant. Scientific evidence, as discussed above, calls into question those assumptions.

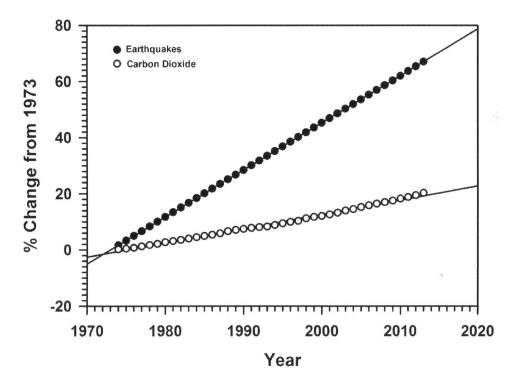

Figure 1.6 Percent annual change, relative to 1973, of earthquakes, magnitude ≥6, calculated from regression line in Figure 1.1 and of National Atmospheric and Oceanic Administration (NOAA) published atmospheric CO_2 values. This figure shows that earthquakes of magnitude ≥ 6 are increasing at a greater rate than atmospheric CO_2.

In Figure 1.6, I present the average percent annual increase in number of earthquakes of magnitude ≥6, relative to 1973. In the same figure, I show the percent annual increase in atmospheric CO_2, also relative to 1973. This figure shows that the rate of increase in number of earthquakes of magnitude ≥6 is greater than the rate of increase in atmospheric CO_2.

Earthquakes are frequently associated with volcanic activity, presumably powered by mantle decompression thermal tsunami [26], which can heat oceans; likewise, seawater heating may arise from variable georeactor heat output, such as suggested to account for the periodicity and synchronicity discovered in hotspot lava outpourings from Iceland and the Hawaiian Islands [37, 38]. Earth-heat variability leading to seawater temperature variability, I posit, offers a logical, causally related explanation for the 800,000 year correlation between seawater temperature and atmospheric CO_2, shown in Figure 1.5 [43].

To reiterate: The present day, now-ongoing "scientifically sanctioned" malevolent agenda, referred to as "anthropogenic (human caused) global warming," is based upon science that is just as flawed as that of race hygiene. The flaw is the assumption that human-produced CO_2 annually increases the CO_2 content of the atmosphere causing global warming by the greenhouse effect. The present annual increase in the CO_2 content of the atmosphere, I submit, is caused by increases in Earth-heat production heating seawater.

For more than half a century, geophysicists modeled mantle convection without ever realizing that the underlying assumptions were false. Mantle convection is physically impossible. The mantle is 62% denser (heavier) at the bottom than at the top; the small decrease in density due to bottom-heating, <1%, cannot cause bottom-mantle matter to float to the top. Moreover, unlike as previously assumed, the mantle is *not* adiabatic, *i.e.*, without friction, which is indicated by the catastrophic release of stress by earthquakes at depths as great as 660 km. Further, the Rayleigh Number [44] has been misapplied to wrongly justify mantle convection [25, 30, 31, 45]. For more than half a century, vast resources have been wasted on mantle convection model making with no scientific gain. But, it has been a benign waste. Benign cannot be said about "climate change" modeling activities, which are all based upon the expressed or tacit false assumption that Earth-heat production is constant: That flawed science is now being hijacked to serve political, financial, and self-aggrandizement agendas.

While academics debate the merits [8], the United States Government is already secretly engaged in engineering Earth: In a blog updated on June 1, 2014 and published on huffingtonpost.com, Bill Chameides, hardly a "conspiracy theorist" but the Dean of Duke University's Nicholas School of the Environment, writes:

A heavily redacted copy of a classified report titled "America Cools Down on Climate" (ACDC)... code-named "Rainmaker" ... obtained by TheGreenGrok .... outlines the audacious plan to use commercial air traffic to mitigate the growing impacts of climate change across the United States....The ACDC plan... "manipulate[s] the jet stream to a more southern trajectory during wintertime to create a polar vortex that would spawn intense snowstorms".... "Americans will suffer this winter," the report predicts, "but consumers will be thankful next year when the price of a Coke drops thanks to lowered high fructose corn syrup prices spurred by a bumper corn crop"....Key to the program were apparently specially designed flexible cords made out of nanofibers....but it worked....Last winter the United States saw...the coldest winter in 20 years and snowfall breaking records in many areas of the Northeast and Midwest.

That activity was undertaken without public disclosure, understanding or informed consent. Apparently, there was little appreciation or concern for the potential health risks involved, especially to airline passengers who breathe the local air. Health risks, I submit, should be of paramount concern; no one knows whether the "flexible cords made out of nanofibers" lodge in the lungs as fibers of asbestos do, which are known to cause cancer (mesothelioma). Moreover, in this instance potential health risks are international; jet streams do not recognize political boundaries.

From the *de facto* admission of the above instance of geoengineering, one can reasonably infer that another operation, cloaked in secrecy, is ongoing, namely, spraying highly reflective nanoparticles into the atmosphere to reflect sunlight back into space to compensate for alleged anthropogenic global warming. Popularly referred to as "chemtrails", there is much information on the Internet and in books, *e.g.*, [46], but virtually no public admission, no understanding, no academic investigations, and no informed consent. Instead, there appears to be a systematic pattern of disinformation, efforts to brand concerned observers with the pejorative moniker, "conspiracy theorists," and to falsely imply that chemtrails are actually "contrails," the formation of water/ice condensation occasionally created by high-altitude aircraft [47].

Let me add the words of a humble naturalist and scientist practiced in observation. I have lived in the same house since 1977 and viewed the same area of the sky nearly every day. After the morning marine layer burns off, the sky in San Diego, California, USA, is often cloudless; rain is infrequent here. Over the years, I have seen on the average less than a single contrail a year, and always at great distance. Recently, though, sometimes for several days in a row, the sky is filled with chemtrails. Figures 1.7, 1.8, and 1.9, August 24, 2014 photographs of the area of sky I regularly observe, show chemtrails, not contrails

Figure 1.7 Chemtrails crossing. Note the more dispersed chemtrail terminated as the spray cut off in an environment where chemtrails continued to form, which shows these are *not* H2O-based contrails.

The spray from one aircraft abruptly ceased in a portion of the sky where other chemtrails were being made – not a contrail occurrence. The wispy "clouds" in the background are earlier, dispersed chemtrails; eventually they will form a white haze in the sky unlike contrails, which are made of H_2O. These photographs were taken early in the spraying; later in the day much of the blue sky was obscured.

To my knowledge there is no official acknowledgement of the chemtrails program, which reportedly occurs throughout the United States and over some Western European nations as well. Individuals have had post-chemtrail rainwater samples analyzed; these have been found to contain excess levels of aluminum and barium [46]. Apparently, there have been no systematic studies of health risks: One can reasonably expect adverse health consequences for children, the elderly, and those with compromised immune or respiratory systems. The long-term effects on the human population and on our planet's biota are wholly unknown. These potentially-toxic chemtrail activities are being undertaken on the basis of very flawed and agenda-driven science. Aggressive actions are being undertaken by the United States and some Western European Governments in secret that include spraying vast quantities of high-reflectance chemicals into the atmosphere over cities and elsewhere, that form chemtrails, for the purpose of counteracting alleged anthropogenic greenhouse global warming. No one knows what potentially devastating health and environmental crises might result, including potentially triggering the onset of an ice age which would lead to wide-spread global starvation as it did during the "little ice age," ca. 1650-1850 [48].

Figure 1.8 Photo taken 12 minutes later than the photo shown in Figure 1.6. The "clouds" in the background are previously dispersed chemtrails.

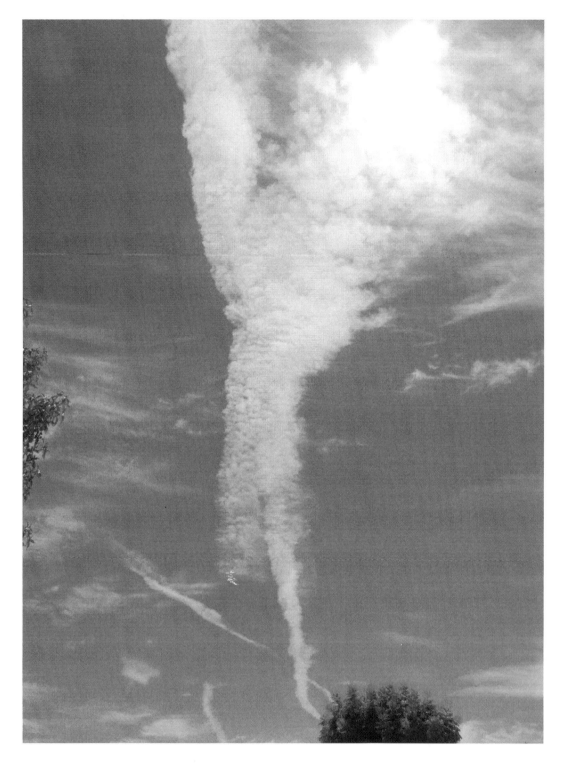

Figure 1.9 Photo of the same chemtrails shown in Figure 1.6, but taken 18 minutes later.

Figure 1.10 November 23, 2014 photo of chemtrails and dispersed chemtrails nearly obscuring the otherwise clear blue San Diego sky with toxic chemicals. These fine particles don't simply stay in the upper atmosphere; some are brought downward by air currents, some slowly settle out, or are brought down with rain. A reasonable person might worry about the health effects on the elderly, those with compromised immune or respiratory systems, aircraft passengers breathing the local air, children playing outdoors, and, especially, the yet to be born children within their mothers' womb. Is the chemical spraying responsible in part for the observed increase of autism in children? No research has been done on the consequences. Later in the day, the same portion of the blue sky was "overcast," even more obscured by the chemical spray.

Liars should never be trusted, whether they are elected officials, political candidates, or global warming activists. On the evening of January 30, 1933, Hitler had only been Chancellor for a few hours when Herman Goering addressed the German people by radio and announced: *"A new chapter opens today and in this chapter liberty and honor will constitute the very basis of the new State."* Not long thereafter, Hitler seized the labor unions, their bank accounts, and pension assets; the very labor unions that had aided in his victory [49]. Left unabated, what new possibilities lay in store for the continual, secret aerial spraying of toxic chemicals over major cities? As proven time and time again by the Nazis, scientists engaged in a malevolent political agenda have no abiding concern for humanity. U. S. President Barack Obama's top science advisor, John Holdren, in 1977 advocated:

Adding a sterilant to drinking water or staple foods ... it must be uniformly effective, despite widely varying doses received by individuals, and ... it must have no effect on ... pets, or livestock" [50].

Note: The "total number of people sterilized during the Nazi period ... is probably nearer 400,000" [1].

I count myself among those who express concern for preserving the environment of our planet and for improving the quality of all life. Certainly many human activities are major, negative, contributive factors, a classic example being the farming activities in the mid-west that stripped the buffalo grass ground cover which led to the 1930s "dust bowl." Covering the surfaces of our planet with cement cannot be without consequences; nor can littering the oceans with plastic waste. Scientists can make a positive difference: Efforts by responsible scientists, for example, have recognized and abated global problems, such as acid rain [51] and destruction of the ozone layer by chlorofluorocarbons [52]. But today, science is in serious trouble. Universities have been lax in their standards, graduating scientists who do not know what is not known, *i.e.*, the limits of knowledge, and who make models based on assumptions instead of making discoveries. Science has been seriously compromised and corrupted by a malevolent political agenda, and will take a long time to recover.

When I realized in 1979 that the Earth's inner core was not necessarily partially crystallized iron metal as had been taught and thought since 1940, instead of discussion, debate, and testing, I was "excommunicated." Without government support, and, consequently, without having to lie to get that support, I was free to make scientific discoveries, constrained only by the necessity of being entirely truthful. I published my discoveries and insights in world-class scientific journals, such as *Proceedings of the Royal Society of London*, *Proceedings of the National Academy of Sciences (USA)*, and *Current Science*, which is published in association with the Indian Academy of Sciences. Over a period of thirty five years, I presented a new understanding of geoscience that, if true, has far-reaching and fundamental implications and which I describe in this book. In this chapter I have introduced a fundamentally different way of explaining the increasing CO_2 content of the atmosphere, not as an anthropogenic contribution with deleterious implications, but as a consequence of Earth-heat raising the temperature of the oceans and decreasing the solubility of CO_2 in the oceans. As some measure of support, I have shown that earthquakes of magnitude ≥ 6 are increasing at a greater rate than atmospheric CO_2 (Figure 1.5) and that there are far-from-negligible connections between earthquake activity, atmospheric CO_2, and global mean seawater temperature.

Life on Earth exists in a state of ultra-delicate balance, poised as a fragile membrane between powerful, dynamic energy sources that are linked by electromagnetic induction and yet held in opposition by the georeactor's magnetic field [20, 24, 27]. While serious efforts are being made by many to understand the Sun's dynamics and interactions with Earth, I seem to be nearly alone in seeking to understand the georeactor, the potential limits of its lifetime [34], its influence on our planet, its inductive connection to changes in the solar wind [25], and the consequences of Earth's early origin as a Jupiter-like gas giant [20].

There have been major species-extinction events in the geological past, continent-splitting is still ongoing, the geomagnetic field is rapidly changing, the occurrence of earthquakes of magnitude ≥ 6 is increasing; ours is a dynamic, still-decompressing planet, one whose manifold connected properties, I will attempt to show you, are far more economically and elegantly explained by the heterodox (but more encapsulating of available evidence) thesis of planetary decompression than the static, un-reexamined and orthodox picture of standard geoscience.

The role of the geoscientist should be to understand the natural processes taking place, both for the sake of knowledge (*scientia*) itself and so that people can be prepared for whatever is to come. It is a naïve and foolish notion that humans can engineer Earth to the benefit of humanity. Yes, humans have the technology to work major changes to our planet, but with unforeseen and potentially devastating consequences. Also, if America and Western European nations, in their apparent profound arrogance, endanger the climate, environment, and wellbeing of others in the world, it may be naïve and foolish to expect no retaliation. I would prefer that destruction not be our destiny nor that scientists, wittingly or not, be its authors along with those whom they serve.

As for what lies ahead in this book, permit me to make a few suggestions. Often individuals when encountering a new concept will attempt subconsciously to paste it into the fabric of their extant understanding, which in the present instance might lead to confusion. Instead, try this: Begin reading with an open mind, as much as possible a *tabula rasa*, a blank slate. Then follow the logical progression of understanding that I lay out. By that approach, you may begin to see where I am coming from, where I am going, and how I am getting there. And, along the way you may begin to see the elegant simplicity and beauty of nature. Those who are scientists and hunger for more may find it in the Appendices, copies of some of my published scientific papers. The non-scientists may stumble over a few technical details, but the big-picture should be clear and along with it a roadmap of the new view and its scientific and other implications.

Chapter 1 References

1. Proctor, R.N., *Racial Hygiene: Medicine Under the Nazis*1988, Cambridge, MA: Harvard University Press.

2. Baur, E., E. Fischer, and F. Lenz, *Grundriß der menschlichen Erblehre und Rassenhygiene (Outline of Human Genetics and Racial Hygiene)*1921, Munich: Julius Friedrich Lehmann.

3. Remy, S.P., *The Heidelberg Myth: The Nazification and Denazification of a German University*2002, Cambridge, Massachusetts: Harvard University Press.

4. Friedlander, H., *The Origins of Nazi Genocide: From Euthanasia to the Final Solution* 1995, Chapel Hill, North Carolina, USA: University of North Carolina Press.

5. Costella, J., ed. *The Climategate Emails.* 2010, The Lavoisier Group: Australia.

6. Solomon, L., *The Deniers: The World Renowned Scientists Who Stood Up Against Global Warming Hysteria, Political Persecution, and Fraud.*2008, Minneapolis, MN: Richard Vigilante Books.

7. Inhofe, J., *The Greatest Hoax: How the Global Warming Conspiracy Threatens Your Future*2012, Washington, DC: WND Books.

8.   Showstack, R., *Scientists debate geoengineering at European Geosciences Union meeting*, in *EOS*2014, American Geophysical Union: Washington, DC.

9.   Spencer, R.W., *The Great Global Warming Blunder: How Mother Nature Fooled the World's Top Climate Scientists* 2010, New york: Encounter Books.

10.   Sussman, B., *Climategate: A Veteran Meteorologist Exposes the Global Warming Scam*2010: WND Books

11.   Revelle, R. and H.E. Suess, *Carbon dioxide exchange between atmosphere and ocean and the question of an increase of atmospheric CO2 during the past decades.* Tellus, 1957, 9: p. 18-27.

12.   Benfield, A.F., *Terrestrial heat flow in Great Britain.* Proc. R. Soc. Lond, 1939. Ser A 173: p. 428-450.

13.   Bullard, E.C., *Heat flow in South Africa.* Proc. R. Soc. Lond, 1939. Ser. A 173: p. 474-502.

14.   Revelle, R. and A.E. Maxwell, *Heat flow through the floor of the eastern North Pacific Ocean.* Nature, 1952. 170: p. 199-200.

15.   Stein, C. and S. Stein, *A model for the global variation in oceanic depth and heat flow with lithospheric age.* Nature, 1992. 359: p. 123-129.

16.   Blackwell, D.D., *The thermal structure of continental crust*, in *The Structure and Physical Properties of the Earth's Crust, Geophysical Monograph 14*, J.G. Heacock, Editor 1971, American Geophysical Union: Washington, DC. p. 169-184.

17.   Kellogg, L.H., B.H. Hager, and R.D. van der Hilst, *Compositional stratification in the deep mantle.* Science, 1999. 283: p. 1881-1884.

18.   Herndon, J.M., *Whole-Earth decompression dynamics.* Current Science, 2005. 89(10): p. 1937-1941.

19.   Herndon, J.M., *Inseparability of science history and discovery.* Hist. Geo Space Sci., 2010. 1: p. 25-41.

20. Herndon, J.M., *New indivisible planetary science paradigm.* Current Science, 2013. 105(4): p. 450-460.

21. Herndon, J.M., *The nickel silicide inner core of the Earth.* Proceedings of the Royal Society of London, 1979. A368: p. 495-500.

22. Herndon, J.M., *Feasibility of a nuclear fission reactor at the center of the Earth as the energy source for the geomagnetic field.* Journal of Geomagagnetism and Geoelectricity, 1993. 45: p. 423-437.

23. Herndon, J.M., *Sub-structure of the inner core of the earth.* Proceedings of the National Academy of Sciences USA, 1996. 93: p. 646-648.

24. Herndon, J.M., *Solar System processes underlying planetary formation, geodynamics, and the georeactor.* Earth, Moon, and Planets, 2006. 99(1): p. 53-99.

25. Herndon, J.M., *Terracentric nuclear fission georeactor: background, basis, feasibility, structure, evidence and geophysical implications.* Current Science, 2014. 106(4): p. 528-541.

26. Herndon, J.M., *Energy for geodynamics: Mantle decompression thermal tsunami.* Current Science, 2006. 90(12): p. 1605-1606.

27. Herndon, J.M., *Nuclear georeactor generation of the earth's geomagnetic field.* Current Science, 2007. 93(11): p. 1485-1487.

28. Herndon, J.M., *Indivisible Earth: Consequences of Earth's Early Formation as a Jupiter-Like Gas Giant*, L. Margulis, Editor 2012, Thinker Media, Inc.

29. Herndon, J.M., *Beyond Plate Tectonics: Consequence of Earth's Early Formation as a Jupiter-Like Gas Giant*, 2012, Thinker Media, Inc.

30. Herndon, J.M., *Origin of mountains and primary initiation of submarine canyons: The consequences of Earth's early formation as a Jupiter-like gas giant.* Current Science, 2012. 102(10): p. 1370-1372.

31. Herndon, J.M., *Geodynamic Basis of Heat Transport in the Earth.* Current Science, 2011. 101(11): p. 1440-1450.

32. Kuczera, B., *Radiogenic Heat Production and the Earth's Heat Balance.* ATW-International Journal for Nuclear Power, 2008. 53(11): p. 674-680.

33. Pollack, H.N., S.J. Hurter, and J.R. Johnson, *Heat flow from the Earth's interior: Analysis of the global data set.* Rev. Geophys., 1993. 31(3): p. 267-280.

34. Herndon, J.M., *Nuclear georeactor origin of oceanic basalt $^3He/^4He$, evidence, and implications.* Proceedings of the National Academy of Sciences USA, 2003. 100(6): p. 3047-3050.

35. Bijwaard, H. and W. Spakman, *Tomographic evidence for a narrow whole mantle plume below Iceland.* Earth Planet. Sci. Lett., 1999. 166: p. 121-126.

36. Nataf, H.-C., *Seismic Imaging of Mantle Plumes.* Ann. Rev. Earth Planet. Sci., 2000. 28: p. 391-417.

37. Mjelde, R. and J.I. Faleide, *Variation of Icelandic and Hawaiian magmatism: evidence for co-pulsation of mantle plumes?* Mar. Geophys. Res., 2009. 30: p. 61-72.

38. Mjelde, R., P. Wessel, and D. Müller, *Global pulsations of intraplate magmatism through the Cenozoic.* Lithosphere, 2010. 2(5): p. 361-376.

39. Herndon, J.M., *Impact of recent discoveries on petroleum and natural gas exploration: Emphasis on India.* Current Science, 2010. 98(6): p. 772-779.

40. Shearer, P.M. and P.B. Stark, *Global risk of big earthquakes has not recently increased.* Proc. Nat. Acad. Sci. USA, 2012. 109(3): p. 717-721.

41. Al-Anezi, K. and N. Hilal, *Scale formation in desalination plants: effect of carbon dioxide.* Desalination, 2007. 204: p. 385-402.

42. Weiss, R.F., *Carbon dioxide in water and seawater: the solubility of a non-ideal gas.* Mar. Chem., 1974. 2: p. 203-215.

43. Herndon, J.M., *Variable Earth-heat production as the basis of global non-anthropogenic climate change.* Submitted to Current Science, September 4, 2014.

44. Rayleigh, L., *On convection currents in a horizontal layer of fluid, when the higher temperature is on the under side.* Phil. Mag., 1916. 32: p. 529-546.

45. Herndon, J.M., *Nature of planetary matter and magnetic field generation in the solar system.* Current Science, 2009. 96(8): p. 1033-1039.

46. Kirby, P.A., *Chemtrails Exposed*, 2012.

47. Oliver, J.E. and T.J. Wood, *Conspiract theories and the paranoid styles of mass opinion.* American Journal of Political Science. doi: 10.1111/ajps.12084 2014

48. Deming, D. *The Coming Ice age.* 2009.

49. Delarue, J., *The Gestaop: A History of Horror*1962, New York: Skyhorse Publishing.

50. Erlich, P.R., A.H. Erlich, and J. Holdren, *Ecoscience: Population, Resources, Environment*1977: Freeman.

51. Likens, G.E., F.H. Bormann, and N.M. Johnson, *Acid rain.* Environment, 1972. 14(2): p. 33-40.

52. Molina, M.J. and F.S. Rowland, *Stratospheric sink for chlorofluoromethanes: Chlorine atom-catalysed destruction of ozone.* Nature, 1974. 249: p. 810-812.

53. BAS, *Science Briefing - Ice Cores and Climate Change,* http://www.antarctica.ac.uk/press/journalists/resources/science/icecorebriefing.php.

54. Jouzel, J., et al., *Orbital and millennial Antarctic climate variability over the last 800,000 years.* Sci., 2007. 317: p. 793-796.

55. Lüthi, D., ey al., *High-resolution carbon dioxide concentration record 650,000-800,000 years before present.* Nature, 2008. 453: p. 379-382.

2. Unexpected Depths and Shifting Surfaces

That man will be very fortunate who, led by some unusual inner light, shall be able to turn from the dark and confused labyrinths within which he might have gone forever wandering with the crowd and becoming ever more entangled. Therefore, in the matter of philosophy, I consider it not very sound to judge a man's opinion by the number of his followers.– *Galileo Galilei, The Assayer (1623)*

Consider the planets of the Solar System (Figure 2.1). The four inner planets are "rocky"; the four outer are "gas giants." Yet, there is good reason to believe that all planets formed from primordial matter of the same composition, matter like that in the photosphere, the outer part, of the Sun.

Figure 2.1 Upper image: Comparison of the relative sizes of the planets of our Solar System [Mercury, Venus, Earth, Mars, Jupiter, Saturn, Uranus, and Neptune]. Note that the distances between these objects in the upper image are not to scale. Lower image: The distance of each planet from the Sun, expressed in Astronomical Units (AU), the distance between Earth and Sun.

In that portion of the Sun, the "gaseous" elements are much more abundant than the "rock-forming" elements. Approximately the same mix of rock-forming elements in the Sun is found to comprise several types of primitive meteorites. In fact, if our "rocky" planet were put back together with its original complement of primordial gas, Earth would be about 300 times as massive, a gas giant almost identical to Jupiter. So, what happened to the inner planets' gases? The answer to this fundamental question sets my work apart from the fifty year old "consensus" view of Solar System formation that developed from an assumption-based model. Moreover, it leads to fundamentally new insights, not only on Solar System formation, but on the origin, composition, geomagnetism, and dynamics of our own planet.

Since the first hypothesis about the origin of the Sun and the planets was advanced in the latter half of the 18th Century by Immanuel Kant and modified later by Pierre-Simon de Laplace, various ideas have been put forward. Generally, concepts of planetary formation fall into one of two categories that involve either:

- **Condensation from gas at high-pressures, hundreds to thousands times the pressure of Earth's present atmosphere; or**
- **Condensation from gas at very low-pressures, much, much less than one atmosphere.**

Beginning in the 1940s, the idea of condensation from within a giant gaseous protoplanet, with pressures of hundreds to thousands of atmospheres was discussed [1-3]. But then in 1963 Cameron [4] made a model based upon the assumption that primordial dust condensed from a gas of solar composition at a pressure of about one ten-thousandth of an atmosphere. More models were made based upon the assumption that the condensed dust then gathered into progressively larger particles, then into rocks, ultimately becoming planetesimals, which finally gathered to become planets [5, 6]. This modern version of the circa 1900 Chamberlain-Moulton "planetesimal hypothesis" became the popular, but problematic, part of the NASA storyline that is often referred to as the "standard model of solar system formation." For the inner or terrestrial planets, this model necessitates the additional assumption of a "magma ocean," whole-planet melting in order that matter of initially uniform composition would melt so that iron could drain to the planet's center to form the core. Finally, it necessitates a further assumption to explain what happened to the gases that were originally associated with the inner planets: that during planetary formation the inner portion of the Solar System was too warm for ice/gas condensation, but beyond the hypothetical "frost line," between Mars and Jupiter, ices and gases could condense, and gas giant planets could form, or at least that was the assumption. (Figure 2.2)

Figure 2.2 NASA artist's representation of the solar nebula produced to show the "frost line" which I modified by adding the ellipse to show more clearly that hypothetical line.

The "standard model of solar system formation" is so deeply ingrained in planetary scientists' thinking that it was applied without question to other planetary systems. So, how were exoplanet gas giants explained that were observed as close (or closer) to their star than Earth is from the Sun [7]? Rather than questioning the validity of that model, another *ad hoc* assumption was added, namely, the rather wild idea of planet migration. Close-to-star gas giants were assumed to have formed at Jupiter-like distances and then migrated inward to reside in their present orbits. On NASA's website, one of the "Big Questions" is "Why is it that our Jupiter and Saturn did not migrate inward?" But then, the *ad hoc* planet migration model thinking migrated to models of our own Solar System:

> **In our solar system, the "Grand Tack" model has Jupiter migrating closer to the Sun, approaching 1.5 astronomical units (AU) from the Sun before the presence of Saturn caused Jupiter to reverse direction back out to its current location at 5.2 AU [8].**

Did anyone in the planetary science community ever stop and ask the question, "What's wrong with this picture?" I did that by using thermodynamic considerations to question the nature of the compounds that are expected to condense from a gas of solar composition at a pressure of one ten-thousandth of an atmosphere. At low pressures in that medium, condensation takes place at low temperatures in an oxidizing environment. I was able to show [9, 10] by thermodynamic calculations that the condensate of primordial matter at those very low pressures would be oxidized, like the minerals of the Orgueil C1/CI meteorite wherein virtually all elements are combined with oxygen. Instead of condensed iron metal, iron oxides would form, for example, magnetite, Fe_3O_4. The inner planets all have massive cores, as is known from their high relative densities. But, low-pressure, low-temperature condensate from primordial solar matter, if accumulated to form planets, would form planets with insufficiently massive cores; that would be contrary to observations of massive-core inner planets [10, 11]. In other words, the "standard model of solar system formation" does not jibe with the facts of inner planet cores. But who in the university-government complex acknowledges that fundamental inconsistency?

Nevertheless we want to know: What happened to the inner planets' gases?

In 1944, in the darkest days of Nazi Germany, Arnold Eucken, the director of the Göttingen Physicochemical Institute, had a profound insight that would shed light on solar system formation decades later [1]. (Figure 2.3) Eucken understood well the power of thermodynamic calculations. With thermodynamic calculations, for example, it is possible to calculate the composition of rocket exhaust or the compounds that will condense from primordial matter under a given set of circumstances. Eucken used thermodynamic calculations to determine Earth's formation, the temperature-sequence of compounds that condense within a giant gaseous protoplanet at high pressures, pressures hundreds to thousands of times that of our atmosphere.

Figure 2.3 Arnold Eucken (1884-1950) inset on cover of his 1944 paper.

Inside the high-pressure giant gaseous protoplanet, Eucken found, iron metal, and the elements that dissolve in it, condenses first as a liquid, before the rocky compounds that comprise Earth's mantle had even begun to condense. In other words, under those conditions while all other elements were gases, molten iron metal condensed and began forming the Earth's core. Imagine rain, not water rain, but raindrops of molten iron infalling and collecting at the gravitational center to form the Earth's core; later the rock-forming elements condensed and gathered atop the already formed core. Thus, there is no need to assume whole-planet melting for iron to drain down to form the core. Eucken had a great idea, and he justified it with calculations that show how the matter involved behaves. But what was lacking was a way to tie directly his idea and calculations to our planet. That's what I did.

The great majority of meteorites observed landing on Earth are called ordinary chondrites. In the 1940s, when geoscientists collectively decided that Earth resembles an ordinary chondrite meteorite, they ignored enstatite chondrites, which are rarely observed landing. The enstatite chondrites are not only rare, but formed under conditions that seriously limited their oxygen content. Said another way, their minerals are "highly reduced." This essentially means that the minerals contain less oxygen than if they had formed in a more oxidizing environment, such as presently exists on Earth's surface, or in space where other types of chondrites originated. The cause of this extreme state of reduction was not at all understood until Hans E. Suess and I discovered a connection between high-pressure condensation and state of reduction [10, 12].

In the mid-1970s, I investigated enstatite chondrites. By similar, extended thermodynamic calculations, I verified Eucken's results and, while doing so, deduced that oxygen-starved, highly-reduced matter characteristic of enstatite chondrites and, by inference, also the Earth's interior condensed at high temperatures and high pressures from primordial gas under circumstances that isolated the condensate from further reaction with the gas at low temperatures [10, 12]. Moreover, I was able to show that the relative masses of various minerals in an enstatite chondrite, observed through the microscope, match quite precisely the relative masses of corresponding components of the Earth's interior, determined from earthquake data [13-16]. I went a step further, though, and posited that Earth's complete condensation formed a gas giant planet virtually identical in mass to Jupiter [11, 17, 18]. Meanwhile and to this day, scientists of the university-government complex engage in the wildly improbable *ad hoc* "gas giant planet migration" models to explain Jupiter-sized extrasolar planets at Earth-like distances from their stars.

My concept of Earth formation by raining-out from within a giant gaseous protoplanet and forming a fully condensed gas giant planet, like Jupiter, is fundamentally different from the "standard model of solar system formation" in essentially all aspects. For example, it is not necessary to hypothesize whole-planet melting; core formation occurs naturally on the basis of relative volatility during planetary formation with the least volatile major element, liquid iron, condensing and accumulating first before the more volatile substances even condense. But there are many other differences, as I describe in the scientific literature and in this book. But, back to our question: What happened to the inner planets' gases?

This is what happened to the inner planets' gases: A brief period of violent activity, the T-Tauri phase, occurs during the early stages of star formation with grand eruptions and super-intense "solar-wind." The Hubble Space Telescope image of an erupting binary T-Tauri star can be seen in Figure 2.4.

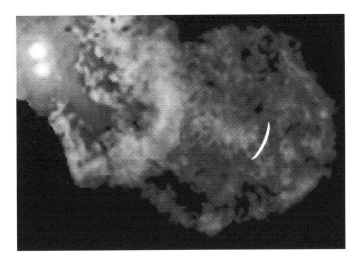

Figure 2.4 Hubble Space Telescope image of binary star XZ-Tauri in 2000 showing a T-Tauri phase outburst. The white crescent label shows the position of the leading edge of that plume in 1995, indicating a leading-edge advance of 130 AU in five years. T-Tauri eruptions are observed in newly formed stars. Such eruptions from our nearly-formed Sun, I submit, stripped the primordial gases from the inner four planets of our Solar System.

The white crescent shows the leading edge of the plume from a five-year earlier observation. In just five years, the plume edge moved 130 astronomical units or AU, a distance 130 times that from the Sun to Earth. T-Tauri outbursts by our young Sun, I posited, stripped gas from the inner four planets. That mechanism not only explains what happened to the primordial gases originally associated with the inner planets, but it provides new insight into the long-standing question of why terrestrial planets, although having formed from essentially identical matter, are so different, especially in their surface behavior. As we shall see, the differences among the inner planets arise primarily because they experienced different degrees of compression from having different size gas/ice shells. Also, as we shall see, the origin of matter in the asteroid belt can for the first time be described in a logical and causally consistent manner [19] by the combination of stripped matter from Mercury's protoplanet fusing with mater from the outer portion of the Solar System. First, though, back to Earth.

During the entire period of the development of the theory of plate tectonics – the explanation for Wegener's continental drift –, the planetesimal theory dominated scientific thinking. The planetesimal theory is the "standard model of solar system formation" – the idea that planets form from dust grains that collided and stuck together, becoming progressively larger bodies, and finally planets. In that framework, geoscience progressed inductively, just as Wegener described. But there was confusion; the pieces did not seem to fit or work together. And, there was no logical and causally related understanding, for example, as to why about 41% of Earth's surface consists of continental rock, while the balance is ocean-floor basalt. Neither proponents of "continental drift" nor "plate tectonics" ever anticipated Earth's early formation as a Jupiter-like gas giant. But I did. As a consequence of Earth's early formation as a Jupiter-like gas giant, I was able to deduce a picture of Earth's formation and composition quite different from the accepted model. Points of difference include:

- **Earth's internal composition and highly-reduced oxidation state;**
- **Core formation without whole-planet melting;**
- **Powerful new internal energy sources, protoplanetary energy of compression and "georeactor" nuclear fission energy;**
- **Mechanism for heat emplacement at the base of the crust that is responsible for the "geothermal gradient";**
- **Georeactor geomagnetic field generation;**
- **Decompression-driven geodynamics that accounts for the myriad of observations attributed to plate tectonics without requiring physically-impossible mantle convection;**
- **The origin of ocean basins;**
- **Fold-mountain formation that does not necessarily require plate collision; and**
- **Primary initiation of fjords and submarine canyons.**

In the following pages I describe all of these consequences and more. For the moment though, the question is in what order? More than thirty years ago, I began the logical progression of understanding with a new concept for the Earth's inner core (Appendix A) which led step by step to everything else. But how many people really relate to the deep, inaccessible interior of the Earth? On the other hand, what happens at the surface is there for all to see, so there we begin – if not at the beginning, then at the surface.

The Earth has existed for about 4½ billion years, but the mapmakers' reasonably complete, reliable representations of her surface were only available after the four voyages of Christopher Columbus between the years 1492 and 1504. In the years surrounding 1570, the Flemish mapmaker, Abraham Ortelius, produced several maps of the world which for the first time began to resemble our modern satellite-imaged, global view of Earth. (Figure 2.5)

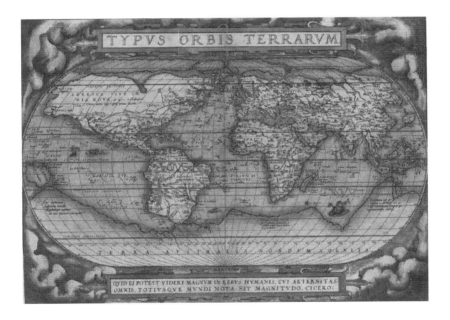

Figure 2.5 Map of the world produced in 1570 by cartographer Abraham Ortelius (1527-1598).

From those maps, Ortelius observed the similarities in the shapes of the eastern coastlines of the Americas and the western coastlines of Europe and Africa. That observation led to his 1596 idea that the Americas were torn away from Europe and Africa by earthquakes and floods. The central idea is that the continents were previously attached and then were separated. This idea appears again and again, albeit with different explanations for the underlying cause. For example, the French prior, François Placet, in 1668, the German theology professor, Theodor Christoph Lilienthal, in 1756, and the French mapmaker, Antonio Snider-Pellegrini, in 1858, each noted the apparent fit of the continent coastlines, as had Ortelius, and each proposed that separation was caused by the biblical events. In 1858, Antonio Snider-Pellegrini published a pair of globe-images showing a close-fitting concentration of continents before and after their separation, which he attributed to a volcanic explosion on the sixth day of Creation [20]. (Figure 2.6)

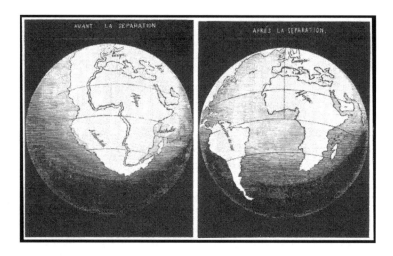

Figure 2.6 The opening of the Atlantic Ocean by Antonio Snider-Pellegrini (1802-1885) from his 1858 book *La Création et ses mystères dévoilés* (*Creation and its Mysteries Unveiled*).

Snider-Pellegrini went further. He noted the similarity of fossil plants in coal beds of North America and Europe and, imagining them together, published a "supercontinent reconstruction." From the coal-bed fossils Pellegrini deduced the continents had been contiguous as recently as about 300 million years ago. Remarkably, Snider-Pellegrini's image is similar to the "Urkontinent" – later "Pangaea" – of Alfred Wegener's continental drift theory, published first as a journal article in 1912 [21], then in 1920 as a book in German, *Die Entstehung der Kontinente und Ozeane* [22], and in English translation as *The Origin of Continents and Oceans* [23]. (Figure 2.7)

In 1912, the German meteorologist, Alfred Wegener first published his idea that about 300 million years ago the continents were joined into one supercontinent, now called Pangaea, that subsequently broke apart with the pieces drifting to their present locations [21]. Although his processors, such as Snider-Pellegrini had similar ideas, Wegener presented considerable evidence in support of his continental drift theory. For example, Wegener pointed to the similarity on opposite sides of the Atlantic of plant and animal fossils, and complementary geological features such as glacial sediments and Carboniferous coal beds. Despite such strong evidence, continental drift nevertheless suffered from the absence of a plausible mechanism and energy source to propel the massive rocky continents.

Figure 2.7 Alfred Wegener (1880-1930) photo ca. 1912.

During the first half of the 20th century, members of the geological establishment generally embraced the idea that Earth was cooling and contracting like a dried apple. Wegener's continental drift theory was not well received; some geologists in fact were downright acrimonious. So, for nearly fifty years Wegener's continental drift languished, although a few prominent geologists, such as Franz Kossmat and Arthur Holmes, were open to the idea.

In 1933, Hilgenberg published a fundamentally different idea about the continents than Alfred Wegener [24]. Hilgenberg imagined, without understanding a reason why, that initially the Earth was smaller in diameter, without oceans, and that the continents fit together like pieces of a jigsaw puzzle and formed a uniform shell of matter covering the entire surface of the planet. Hilgenberg's idea was that the Earth for unknown reasons subsequently expanded. This became the basis for Earth expansion theory, which never developed as a coherent theory as it:

- **Lacked a logical reason for Earth's smaller size.**
- **Had no adequate energy source for expansion.**
- **Offered no explanation for fold-mountain ranges.**

Moreover, the Earth expansion theory was compromised by a plethora of *ad hoc* explanations, without a fundamental basis, such as the generation of new matter within the earth.

In 1931, Holmes suggested the idea of mantle convection as motive force for continental drift [25]; mantle convection was later adopted in the mid-20th century revision of Wegener's theory that became known as plate tectonics theory. In Holmes' mantle convection idea, the rocky part of the Earth is assumed to circulate in great loops, like endless conveyer belts, dragging the continents along. (Figure 2.8)

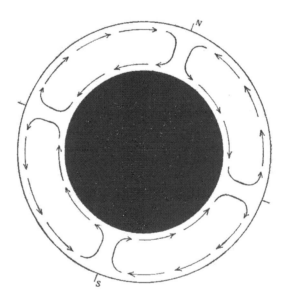

Figure 2.8 Schematic representation of mantle convection published in 1931 by Arthur Holmes (1890-1965). Reproduced with permission of the Geological Society of Glasgow.

Before World War II relatively little was known about the land beneath the oceans. Submarine mountains had been discovered in the middle of the Atlantic Ocean, but the extent of the mid-ocean ridges, as they are now called, only became apparent from echo-sounding measurements made after that war: Now it is known that the global system of mid-ocean ridges encircles the Earth, like the stitching on a baseball, and forms a global chain of active volcanoes. (Figure 2.9)

Before World War II relatively little was known about the land beneath the oceans. Submarine mountains had been discovered in the middle of the Atlantic Ocean, but the extent of the mid-ocean ridges, as they are now called, only became apparent from echo-sounding measurements made after that war: Now it is known that the global system of mid-ocean ridges encircles the Earth, like the stitching on a baseball, and forms a global chain of active volcanoes. (Figure 2.9)

During World War II, Harry H. Hess, Captain of the U.S.S. Cape Johnson, an assault transport, let his vessel's echo-sounding equipment run continuously during long traverses across the Pacific Ocean. Hess was thus able to reveal the profiles of extensive tracts along the Pacific Ocean's floor. It was like looking into a once-hidden abyss and for the first time clearly seeing some of its fantastic features. Hess discovered flat-topped, undersea mountains in the South Pacific that he named guyots. He realized that these were originally volcanic mountains, whose heights over time eroded first to sea level where they existed for a time as the support base for coral atolls. Slowly, the movement of the seafloor to which they were attached submerged them into deeper water, and finally brought them to the edge of a trench into which they would ultimately plunge. Reflecting on these observations, Hess proposed the concept of seafloor spreading, which is the idea that oceanic basalt, the rock comprising seafloors, continually erupts from volcanoes at the mid-ocean ridges, creeps across the ocean bottom, and ultimately plunges into the Earth at trenches that often occur beside the continents [26].

While the seafloors appear to be mobile, the mid-ocean ridge system is a fixed structural feature of Earth. There are other more-or-less fixed structures on the Earth called "hotspots" (or "hot spots"). These are points on the Earth's surface where lava, melted by heat from deep within the Earth, rises to the surface. In 1963, J. Tuzo Wilson envisioned the chain of Hawaiian Islands and the extended, submerged and related chain, the Emperor Seamounts, as having formed sequentially over time as the seafloor passed over a persistent hotspot that is currently situated beneath the Island of Hawaii. Subsequent investigations supported that concept by showing that the islands are progressively older the further they stretch along the chain away from the Island of Hawaii [27]. (Figure 2.10)

Figure 2.9 Modern representation of the topography of Earth's surface. Note the mid-ocean ridges that encircle the globe like stitching on a baseball. As will become evident, these features and more are understandable logically and causally as the consequence of Earth's early formation as a Jupiter-like gas giant.

Figure 2.10 U.S.G.S. representation of the Hawaiian Islands. The oldest island is on the left, progressing sequentially in time to the youngest island on the right as a consequence of ocean crust moving over a fixed, persistent hotspot.

Both the guyots and the hotspot island chains provide compelling evidence of at least part of Hess' seafloor spreading concept. But really striking support came from the results of magnetic surveys of the seafloor first made in the 1960s. The Earth's magnetic field reverses polarity from time to time on an irregular basis with an average time between reversals of about 200,000 years; the last magnetic reversal occurred about 750,000 years ago. As erupted molten oceanic basalt cools, its magnetic minerals act like a magnetic memory and record the direction of the geomagnetic field at the time of cooling. Magnetic surveys of the seafloor reveal a pattern of parallel stripes, indicating the alternating "normal" and "reversed" magnetization, symmetric on either side of mid-ocean ridges.

The observed pattern of magnetic stripes across the ocean floor reflects a series of geomagnetic reversals, recorded as new seafloor cools and moves away from the mid-ocean ridge. Cores taken from the seafloor and dated indicate progressively older ages at progressively greater distances from the mid-ocean ridge. (Figure 2.11)

During the late 1960s and into the 1970s, plate tectonics theory was developed from seafloor spreading and incorporated that concept; seafloor is produced by volcanos at mid-ocean ridges, moves across the expanse of the ocean, disappears into trenches where it is thought to be re-circulated through the mantle by convection, like great conveyer belts.

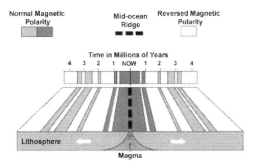

Figure 2.11 Schematic representation of seafloor magnetic stripes, symmetric on each side of the mid-ocean ridge, produced as erupted basalt moves progressively away from the ridge.

The idea behind plate tectonics theory is that the surface of the Earth is comprised of seven major plates and a larger number of minor plates, which move as units being pushed or pulled by assumed convection currents in the mantle. (Figure 2.12)

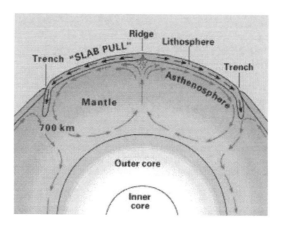

Figure 2.12 U.S.G.S. representation showing the role of assumed mantle convection in plate tectonics theory.

Plate tectonics theory appears to explain many of Earth's surface features, particularly those beneath the oceans. The magnetic stripes and the seafloor core dates might appear to be convincing evidence of plate tectonics theory. But does that mean plate tectonics theory is totally correct? Not necessarily. Let's look deeper and critically.

So, what's wrong with plate tectonics? What's wrong with plate tectonics is the same thing that's wrong with Hess' concept of seafloor spreading. Not the visible part, which seems to explain a lot, but the invisible part, the part that is assumed in order to make everything seem to work within the framework of current understanding, which is not correct at all. What's wrong with the idea introduced by Hess to explain seafloor spreading is the assumption of mantle convection. Plate tectonics theory adopted the assumption of mantle convection from seafloor spreading. And yet now, after more than four decades of research, there is still no unambiguous evidence that mantle convection actually exists.

What is convection? Heat a pot of water on the stove top. Before it starts to boil, the water begins to circulate from bottom to top and from top to bottom. This is called convection and it can be better observed by adding a few tea leaves, celery seeds, or the like, which are carried along by the circulation of water. Convection occurs because heat at the bottom causes the water to expand a bit, becoming lighter, less dense, than the cooler water at the top. Heavy on the top and lighter on the bottom is an unstable, top-heavy arrangement that attempts to regain stability by fluid motions thus giving rise to the observed circulation. This is thermal convection.

So, why not mantle convection? For fifty years, since the idea of seafloor spreading and even longer going back to Holmes in the 1930s, the scientific community has assumed that convection "must" exist in the mantle, even though the mantle is solid, not liquid. Untold millions of dollars have been spent on modeling convection in the Earth's mantle without ever asking whether it's possible in the first place. But I did, I asked this question, and found that there is a serious problem with the idea of mantle convection [16].

Because of the weight above, the mantle is compressed. The bottom of the mantle is about 62% denser (heavier) than the top of the mantle. (Figure 2.13) The very small decrease in density, less than 1%, caused by thermal expansion is not enough to make a "parcel" of bottom-mantle material light enough to float to the top. Sometimes in mantle convection models the bottom-heavy prohibition is obviated by assuming that the solid rock mantle behaves as an ideal gas, "adiabatic," which is to say without viscous losses, without friction. But the mantle is not without friction, as indicated by earthquakes at depths as great as 660 km. The adiabatic assumption is simply wrong: the mantle is not adiabatic. Moreover, the Rayleigh Number, designed to predict convection in a non-compressed thin film, has been repeatedly misapplied to wrongly justify physically impossible mantle convection.

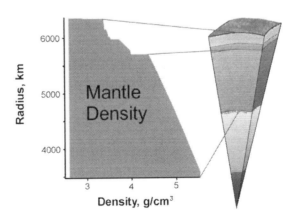

Figure 2.13 The density of the Earth's mantle, showing that the bottom of the mantle is about 62% more dense (heavier) than the top. In other words, the mantle is bottom-heavy, whereas thermal convection occurs for a fluid that is top-heavy.

So, does the circumstance that mantle convection is physically impossible mean that plate tectonics is totally wrong? No. The correctness of the plethora of observations attributed to plate tectonics is not diminished, at least in the majority of cases. The absence of mantle convection just means that a more fundamental understanding of Earth's behavior than plate tectonics is required. A more fundamental understanding of Earth's behavior, though, must also explain the plethora of observations attributed to plate tectonics without invoking mantle convection.

Chapter 2 References

1. Eucken, A., *Physikalisch-chemische Betrachtungen ueber die frueheste Entwicklungsgeschichte der Erde.* Nachr. Akad. Wiss. Goettingen, Math.-Kl., 1944: p. 1-25.

2. Kuiper, G.P., *On the origin of the Solar System.* Proc. Nat. Acad. Sci. USA, 1951. 37: p. 1-14.

3. Kuiper, G.P., *On the evolution of the protoplanets.* Proc. Nat. Acad. Sci. USA, 1951. 37: p. 383-393.

4. Cameron, A.G.W., *Formation of the solar nebula.* Icarus, 1963. 1: p. 339-342.

5. Goldrich, P. and W.R. Ward, *The formation of planetesimals.* Astrophys J., 1973. 183(3): p. 1051-1061.

6. Wetherill, G.W., *Formation of the terrestrial planets.* Ann. Rev. Astron. Astrophys., 1980. 18: p. 77-113.

7. Seager, S. and D. Deming, *Exoplanet Atmospheres.* Ann. Rev. Astron. Astrophys., 2010. 48: p. 631-672.

8. Pierens, A. and S.N. Raymond, *Two phase, inward-then-outward migration of Jupiter and Saturn in the gaseous solar nebula.* Astron. Astrophys., 2011. 533(A131): p. 14.

9.   Herndon, J.M., *Reevaporation of condensed matter during the formation of the solar system.* Proceedings of the Royal Society of London, 1978. A363: p. 283-288.

10.   Herndon, J.M., *Solar System processes underlying planetary formation, geodynamics, and the georeactor.* Earth, Moon, and Planets, 2006. 99(1): p. 53-99.

11.   Herndon, J.M., *Nature of planetary matter and magnetic field generation in the solar system.* Current Science, 2009. 96(8): p. 1033-1039.

12.   Herndon, J.M. and H.E. Suess, *Can enstatite meteorites form from a nebula of solar composition?* Geochim. Cosmochim. Acta, 1976. 40: p. 395-399.

13.   Herndon, J.M., *The chemical composition of the interior shells of the Earth.* Proceedings of the Royal Society of London, 1980. A372: p. 149-154.

14.   Herndon, J.M., *Nuclear fission reactors as energy sources for the giant outer planets.* Naturwissenschaften, 1992. 79: p. 7-14.

15.   Herndon, J.M., *Composition of the deep interior of the earth: divergent geophysical development with fundamentally different geophysical implications.* Phys. Earth Plan. Inter, 1998. 105: p. 1-4.

16.   Herndon, J.M., *Geodynamic Basis of Heat Transport in the Earth.* Current Science, 2011. 101(11): p. 1440-1450.

17.   Herndon, J.M., *Whole-Earth decompression dynamics.* Current Science, 2005. 89(10): p. 1937-1941.

18.   Herndon, J.M., *Indivisible Earth: Consequences of Earth's Early Formation as a Jupiter-Like Gas Giant*, L. Margulis, Editor 2012, Thinker Media, Inc.

19. Herndon, J.M., *Discovery of fundamental mass ratio relationships of whole-rock chondritic major elements: Implications on ordinary chondrite formation and on planet Mercury's composition.* Current Science, 2007. 93(3): p. 394-398.

20. Snider-Pellegrini, A., *La Création et ses mystères dévoilés ("Creation and its Mysteries Unveiled")*1858, Paris: Franck et Dentu.

21. Wegener, A.L., *Die Entstehung der Kontinente.* Geol. Rundschau, 1912. 3: p. 276-292.

22. Wegener, A., *Die Entstehung der Kontinente und Ozeane.* fourth ed1929, Braunschweig: Friedr. Vieweg & Sohn. 246.

23. Wegener, A., *The Origin of Continents and Oceans.* Translated by John Biram1966, New York: Dover Publications, Inc.

24. Hilgenberg, O.C., *Vom wachsenden Erdball*1933, Berlin: Giessmann and Bartsch. 56.

25. Holmes, A., *Radioactivity and Earth movements.* Trans. geol. Soc. Glasgow 1928-1929, 1931. 18: p. 559-606.

26. Hess, H.H., in *Petrologic Studies: A Volume in Honor of A. F. Buddington*1962, Geological Society of America: Boulder. p. 599.

27. Keller, R.A., et al., *Cretaceous-to-recent record of elevated $^3He/^4He$ along the Hawaiian-Emperor volcanic chain.* Geochemistry, Geophysics, Geosystems, 2004. 5(12): p. 1-10.

3. Whole-Earth Decompression Dynamics (WEDD)

May be the most important paper I have ever edited because it covers Earth-sized phenomena over 4,000,000,000 years. That's a "Big Picture" to consider (& quantitatively!!), – *Lynn Margulis, 2010, comment scribbled in the margin of her edit of Indivisible Earth: Consequences of Earth's Early Formation as a Jupiter-like Gas Giant* [1].

In the previous chapter I suggested that a better (more scientific, if you will) understanding of Earth's behavior needs to account for those many observations attributed to plate tectonics, but without invoking mantle convection, as I have shown. Well, as Jean de La Bruyère said in the 17th century, "ce qui barre la route fait faire du chemin" – what blocks the road makes the way. As it turns out, the impossibility of mantle convection helps drive us to a new, better understanding of Earth's behavior, one which integrates more data points, and which I call Whole-Earth Decompression Dynamics (WEDD). WEDD is, above all, a logical consequence of a coherent picture of Earth's dynamics based on its very likely early existence as a gas giant [2-4].

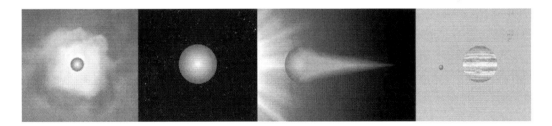

Figure 3.1 Whole-Earth Decompression Dynamics formation of Earth. From left to right, same scale: 1) Earth condensing at the center of its giant gaseous protoplanet; 2) Earth, a fully condensed a gas-giant; 3) Earth's primordial gases stripped away by the Sun's T-Tauri solar eruptions; 4) Earth at the onset of the Hadean eon, compressed to 66% of present diameter shown with Jupiter for size comparison.

The weight of 300 Earth-masses of primordial gases would have gravitationally compressed the non-gases, the original rock-plus-alloy kernel that became the Earth we know, to about 66% of its present diameter. As thermonuclear reactions in the Sun ignited, the violent T-Tauri solar-wind stripped Earth of its Jupiter-like gas envelope. (Figure 3.1) What remained would have been a solid, smaller Earth whose surface consisted entirely of continental rock, without ocean basins. At onset of the Hadean eon, one ancient supercontinent existed: the 100% closed contiguous shell of continental-rock we may well name *Ottland*, in honor of Ott Christoph Hilgenberg, who first conceived its existence [5]. (Figure 3.2)

Figure 3.2 Ott Christoph Hilgenberg (1896-1976). Photographs from 1927 and 1970. Courtesy of daughter, Helge Hilgenberg.

What was Earth like at this point? The core had already formed; in fact, the core was the first part of Earth to form. The crust and perhaps up into the upper mantle was initially quite cold having formed before the Sun ignited. The super-intense T-Tauri solar wind outbursts likely stripped away 300 Earth-masses of primordial gases, which may have cooled the crust even more. There must have been intense bombardment by meteorites and comets in the final stages of Earth formation; these put iron and iron-loving elements, like nickel, in the upper mantle and in the crust. After the primordial gases had been stripped away and the violent T-Tauri phase had ended, water began to collect that was brought to Earth's surface by comets, perhaps the small comets described by Frank [6] in the scientific literature and in his book, *The Big Splash*, that he asserts continue today to bring water to Earth. Volcanic eruptions may have contributed water as well. In the absence of ocean basins, inland seas eventually covered much of the Earth's surface. Oceanic features from that period, such as banded ironstone deposits and pillow basalts that formed from underwater volcanic eruptions may be found in areas that are well within present continent confines.

Meanwhile, deep within the Earth, pressures were building. Occasionally there would be a "blow out." Pressure would force upward a column of matter, from within the mantle at a depth of about 100 km or more. The explosion would puncture a narrow hole a few meters in diameter through all of the overlying rock and then burst forth at the surface producing a funnel shape as wide as 200 meters. These are the diamond-bearing kimberlite pipes. Sporadic, energetic events such as this, though, were just the serenely whispered overture to the global catastrophic violence that would occur again and again, splitting the continental crust, creating a new ocean basin, wreaking havoc, and causing (along with major impact events) widespread extinctions of the species that had evolved on this watery planet. (Figure 3.3)

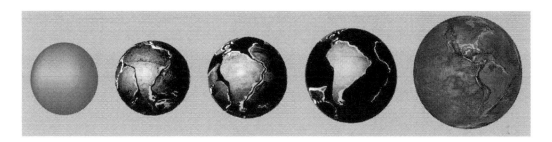

Figure 3.3 Schematic representation of the decompression of Earth from Hadean to present. From left to right, same scale: 1) Earth after T-Tauri removal of gases, 66% of present Earth diameter, fully covered with continental-rock crust; 2), 3), and 4) Formation of primary and secondary decompression cracks that progressively fractured the continental crust and opened ocean basins. Timescale not precisely established. 5) Holocene Earth.

During Earth's formation as a Jupiter-like gas giant, when the rocky kernel was compressed to about 66% of its present diameter, two things happened:

- **The interior got hot due to the heat of compression, with much of the heat presumably lost, and;**
- **Protoplanetary gravitational energy of compression due to the weight of about 300 Earth-masses of primordial gases was stored in the compressed kernel.**

That gravitational energy remained trapped within the rocky kernel, contained by the rigid crust and augmented by the physical properties of crust and mantle material. Then, for the Earth to begin to decompress, two things had to happen:

- **Heat had to be supplied (by natural radioactive decay and nuclear fission) to replace the lost heat of compression; otherwise, decompression would cool the kernel which would impede decompression, and;**
- **Internal pressures had to build sufficiently to produce the force needed to crack the rigid crust.**

From Earth's early origin as a Jupiter-like gas giant, one can deduce the new and fundamental geoscience paradigm that I call Whole-Earth Decompression Dynamics, a perspective which obviates the requirement for physically-impossible mantle convection [2, 4, 7]. In this chapter I provide an initial glimpse into the portion of WEDD that deals with the behavior of Earth's surface. There will be even more about that later, after we discuss Earth's internal composition, energy sources, and modes of heat transport.

After being stripped of its primordial gas envelope, pressures began to build within the compressed, rocky Earth. Eventually, enough pressure developed to begin to crack Earth's rigid crust; much more force is necessary to initiate a crack than for the crack to propagate. No longer contained by the contiguous, rigid crust, Earth decompresses, presumably driven by the stored energy of protoplanetary compression augmented by nuclear fission and radioactive decay energy [8-10]. As Earth decompresses, it increases in diameter and volume. Our planets increasing diameter has important consequences for geomorphology, which I shall now describe.

As the diameter of the Earth increases due to decompression, two fundamental and necessary processes occur at the surface:

- **Cracks form to increase Earth's surface area in response to decompression driven increases in planetary volume [2]; and,**
- **As decompression increases planetary volume, the surface curvature changes, resulting in the formation of mountain chains that are characterized by folding as well as resulting in the initiation of fjords and submarine canyons [11].**

Decompression-driven Increases in Surface Area

Concerning the first process: Cracks in the Earth's surface arise to increase surface area in response to decompression driven increases in volume. According to Whole-Earth Decompression Dynamics there are two types of cracks. *Primary* cracks occur in the crust and are underlain by heat sources capable of extruding lava. Primary decompression cracks are identified with the mid-ocean ridge system. *Secondary* decompression cracks lack underlying heat sources and are typically found along continent margins; they are identifiable as submarine trenches. Secondary decompression cracks are the ultimate repositories for extruded basalt-rock. Basalt-rock, extruded from mid-ocean ridges, traverses the ocean expanse by gravitational creep. Ultimately, in a process of "subduction" that lacks any mantle convection, seafloor basalt, with its carbonate sediment, fills in secondary decompression cracks. Seismically imaged "down-plunging slabs," I submit, are filled-in secondary decompression cracks. (Figure 3.4)

Formation of surface cracks in response to decompression-increased Earth volume is the process by which continent fragmentation and separation takes place over time. This is the process that Alfred Wegener attempted to describe by continental drift theory and it is the process that plate tectonics theory attempts to describe by continents riding atop non-existent mantle plumes. In some cases, plate tectonic observations and interpretations are similar to WEDD, so integration should not be too difficult. In other cases, though, WEDD provides more natural explanations, such as for the origin of mountains characterized by folding, which are attributed to plate collisions in plate tectonics.

Continent-splitting began with the first surface crack in Ottland and is currently on-going along the East African Rift System. (Figure 3.5)

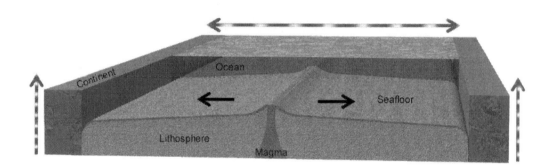

Figure 3.4 Schematic representation of motions related to seafloor spreading according to Whole-Earth Decompression Dynamics. The stippled arrows indicate the directions of whole-Earth decompression movement. Magma is extruded at mid-ocean ridges and creeps by gravity (solid arrows) across the ocean expanse until it falls into and in-fills secondary decompression cracks, observed at continent edges.

Whole-Earth Decompression Dynamics nicely accounts for Earth's well-documented, myriad seafloor related observations, previously ascribed to plate tectonics. Partially in-filled secondary decompression cracks uniquely explain oceanic troughs which are inexplicable by plate tectonics. For those worried that I am making a wholesale dismissal of plate tectonics, don't worry, I am not. Indeed, plate tectonic meanings and terminology are to a great extent preserved in Whole-Earth Decompression Dynamics. For example, *transform plate boundaries* are identical; *divergent plate boundaries* are similar, but with a different driving mechanism; *convergent plate boundaries* likewise are similar, but down-plunging plates neither create oceanic trenches, which are secondary decompression cracks, nor are they recycled through the mantle by conveyer-like mantle convection, and; *Wadati-Benioff earthquake zones* are quite similar, with the possible exception of why mantle melting occurs that is responsible for sometimes associated volcanic eruptions. Many of the plethora of observations, taken to support plate tectonics, support Whole-Earth Decompression Dynamics as well.

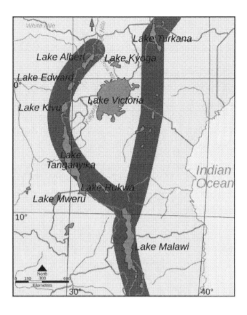

Figure 3.5 The Great Rift Valley, also called the East African Rift System, indicated on this map of Africa, is where continent splitting by primary decompression cracks is currently taking place. Image courtesy of En rouge.

Decompression-driven Changes in Surface Curvature

After the compressed solid crustal shell began to crack, Earth began to decompress, necessitating, in addition to new surface area, changes in its surface curvature. This is easy to envision: Think about placing a circular slice of orange skin atop a larger fruit, for example, a melon, as shown in the demonstration that is the lower portion of Figure 3.6. In this demonstration, the orange represents a section of the ancient compressed Earth crust whereas the melon represents the more decompressed, modern Earth. Clearly, the curvature of the orange slice "continent" is different than the melon's curvature. The orange slice appears to have "extra" surface area contained within its perimeter. So, what might happen naturally that could cause the curvature of the orange slice to conform to the melon's curvature? In other words, what is needed for the skin of the orange slice to match the shape of the melon? One possibility is that stress might cause numerous tears along the edge of the "continent" orange slice. But is that enough? Probably not, as the tears would have to extend to the center of the slice and there would need to be many, many tears. For a continent of the Earth, that would require a vast amount of energy to form so many slits for such distances into the rigid crust. Another, more likely possibility, is that the "extra" surface area might be removed by the formation of tucks. The demonstration image at the lower right of Figure 3.6 illustrates both the formation of tucks and the formation of stress-tears along the perimeter. The upper portion of Figure 3.6 shows a typical example of folded mountain strata. Requisite curvature flattening is accommodated primarily by tucks and secondarily by stress-tears around continent perimeters. The former, I posited, is the basis of the origin of fold-mountains; the latter, the primary initiation of fjords and submarine canyons [11].

Figure 3.6 Mountains characterized by folding (upper), example from the Swiss Alps, are likely the consequence of surface curvature adjustments necessary as the surface of Earth's original, smaller globe had to conform to the later curvature of a larger globe (lower). Upper image courtesy of Woudloper.

Origin of Mountains

Mountain ranges are prominent features of the surface of each continent, for example, the European Alps shown in Figure 3.7. The folded layers of geological strata (Figure 3.6 upper) indicate, as recognized by Eduard Suess [12] in the 19th century, that mountains were not simply pushed up from beneath, which has compounded the mystery of their origin. The age-old problem of mountain formation reduces to this question: How can one account for the "extra" surface area, contained within the continental boundaries, that observations reveal are folded atop other layers on mountains?

Figure 3.7 December 11, 2004 satellite image of the Alps. North of the Alps, clouds cover France, Switzerland, Liechtenstein, Austria, and Slovenia. South of the Alps, clear skies dominate most of the image, leaving the Po River valley and peninsular Italy showing clearly. The Ligurian and Mediterranean Seas are to the left of the Italian peninsula, the Adriatic Sea to the right. NASA image.

During the first half of the 20th century, when many geologists thought the Earth was cooling and contracting, mountain formation was believed to be part of its shriveling up like a dried apple. Then along came plate tectonics, and the story changed to mountain formation by continent collisions. Continental plates, thought to move about freely, riding atop assumed mantle convection cells, supposedly collided causing the continents to become elongated, piling up some of the resulting "extra" surface as mountains. But as mantle convection is physically impossible [4], another more fundamental process must be involved.

The formation of mountains characterized by folding is better understood, I believe, as the inevitable consequence of Earth's decompression from its previous existence as a Jupiter-like gas giant [11]. Earth's continental rock crust necessarily must adjust to decompression-driven changes in surface curvature, as illustrated by the demonstration in Figure 3.6. Earth's surface must buckle, break, and fall over upon itself to adjust to decompression-changed surface curvature; these are the principal agents involved in the formation of fold-mountains.

Not all mountains are formed as I just described: Mountains built up from successive layers of volcanic eruptions are the exception. One prominent example is Mauna Kea in the Hawaiian Islands chain. Presumably, Mons Olympus on Mars is also a volcano-produced mountain.

Fictitious Supercontinent Cycles

Geological literature contains many papers [13-15] dealing with various aspects of "supercontinent cycles", also called "Wilson cycles." This is the idea that before Pangaea there were a series of supercontinents that each formed and then broke apart and separated before colliding again, re-aggregating, and suturing into a new supercontinent in a continuing sequence. Some of these are named Vaalbara, Ur, Kenorland, Columbia, Rodinia, Pannotia, and Pangaea. (Figure 3.8) What are the underlying reasons that necessitate this supercontinent hypothesis?

Mountains have been observed whose ages predate the formation of Pangaea. How can that be if (in plate tectonics) continent collisions are necessary to form mountains? Instead of asking the question "What's wrong with this picture?" Instead of wondering whether there might be a deeper, more fundamental explanation than provided by plate tectonics, geoscientists invented supercontinents with colorful names whose assumed collisions, disruptions, and re-accumulations became part of the supercontinent cycles.

Models of supercontinents engaged in hypothetical Wilson cycles typically make use of problematic paleomagnetic calculations [16]. Moreover, without mantle convection there is no means of supercontinent locomotion. Furthermore, as I realized, fold-mountain formation is a natural consequence of surface curvature adjustments to decompression-increased planetary volume and would not require either mantle convection or plate collisions.

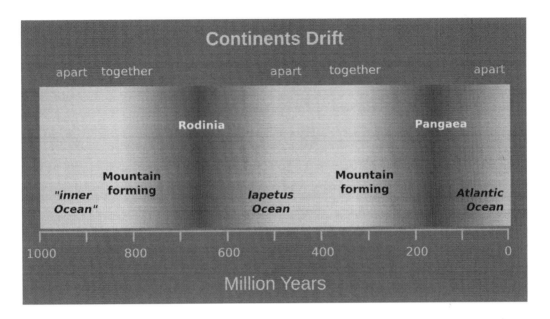

Figure 3.8 Schematic representation of Wilson Cycles, also called Supercontinent Cycles, which attempt to explain by plate tectonics the formation of fold-mountains that predate Pangaea. Image courtesy of Hannes Grobe.

In summary I believe we must question the entire concept of supercontinent cycles. I also believe that the fanciful supercontinent names, including Pangaea, will eventually pass into history. From a WEDD perspective, the challenge for geologists will be to discover the true sequence of fragmentation beginning with Ottland and continuing to the present and to discover the true nature of Earth's surface throughout that progression.

Primary Initiation of Fjords and Submarine Canyons

Almost 150 years ago, Brown [17] wrote: "Intersecting the sea-coast of various portions of the world, more particularly in northern latitudes, are deep, narrow, inlets of the sea, surrounded generally by high precipitous cliffs, and varying in length from two or three miles to one hundred or more, variously known as, "inlets," "canals," "fjords," and even on the western slopes of Scotland as "lochs." The nature of these inlets is everywhere identical, even though existing in widely distant parts of the world, so much so as to suggest a common origin." (Figure 3.9) Here I propose said common origin, which could not have been deduced from 19th and 20th century geodynamics.

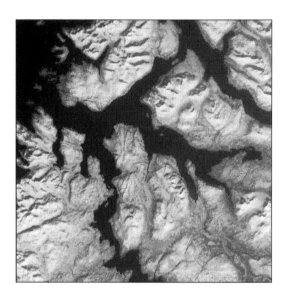

Figure 3.9 Northern Norway fjords: Like dark fingers, cold ocean waters reach deeply into the mountainous coastline of Northern Norway, defining the fjords for which the country is famous. Flanked by snowcapped peaks, some of the fjords are hundreds of meters deep. USGS/NASA image; ASTER data.

The idea that glacial erosion has influenced the formation of fjords has been widely acknowledged since the latter part of the 19th century, although there has been and still is serious debate as to the extent of that influence [18-22]. Some authors have considered erosion by flowing water [23]; others have questioned the interplay of water erosion, glacial activity, and wind erosion [18]. An overriding problem is that most fjords are deeper than current sea level, some having depths in excess of 1 km. In 1913, Gregory [24] maintained that fjords are primarily of tectonic origin with glacial influence being minimal, although his tectonic reasoning seems inadequate to justify such an assertion.

As planetary volume increases, surface curvature must change. Consider a hypothetical circular section of ancient continental crust placed atop a later, more decompressed, larger Earth, as illustrated in Figure 3.6. Note that the section of ancient "crust" (lower portion Figure 3.6, the section of orange skin) contains within its perimeter "excess" surface area that must be "removed" for the ancient section to adjust to the new curvature (in Figure 3.6, the melon curvature). I have suggested that mountains characterized by folding originate as the consequence just such curvature accommodations [11]. As a result of decompression, the "extra" surface area, constrained within a continental perimeter, buckles, breaks, and falls over producing fold-mountains, and thus better conforms to the more-decompressed Earth curvature.

Matter on the surface of a sphere tends to make adjustments to minimize surface energy. Curvature mismatch, as illustrated in Figure 3.6, will cause fold-mountain formation as well as tension fractures along the continent perimeter. These peripheral tension fractures, I posit, are the primary, common origin of fjords.

Tension fractures inevitably form at the edges of continents that are undergoing whole-Earth decompression-caused crustal-curvature adjustments. In the far northern and far southern latitudes, these tension fractures initiate fjords. Glaciation, rather than forming fjords, as widely assumed, preserves them. Observations of fjords, as indicating decompression-driven crustal-curvature adjustments, may be useful for reconstructing geological history and for understanding present geology, especially discoveries made in the ice-obscured, relatively undiscovered areas of Antarctica.

In the middle latitudes tension fractures suffer severe water erosion, yet nevertheless are recognizable as submarine canyons. (Figure 3.10) Submarine canyons are steep-sided valleys cut into the sea floor of the continental slope, sometimes extending well onto the continental shelf. Some submarine canyons are found as extensions to large rivers; however, most of them have no such association. Canyons cutting the continental slopes have been found at depths greater than 2 km below sea level. Many submarine canyons continue as submarine channels across continental rise areas and may extend for hundreds of kilometers. Submarine canyons, for example, occur along roughly 20% of the continental shelf edge between Alaska and the Equator [25].

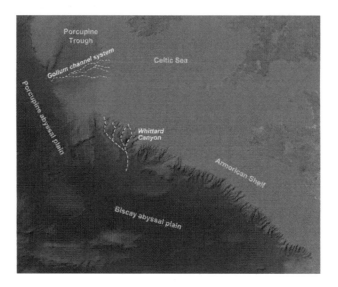

Figure 3.10 Screen capture from NASA WorldWind software of the Atlantic Margin from Southern Ireland to France with the Gollum channel system and Whittard canyon highlighted. Image courtesy of Mikenorton.

Within the plate tectonics paradigm, there is no good explanation for the primary initiation of either fjords or submarine canyons: Subsequent fluvial erosion is widely recognized, but how did submarine canyons begin to form? In Whole-Earth Decompression Dynamics, on the other hand, Earth's early formation as a Jupiter-like gas giant provides a basis for understanding the geomorphology of Earth's surface as a consequence of decompression-driven increases in planetary volume without the necessity of making *ad hoc* postulates, such as plate collisions and supercontinent cycles to explain fold-mountains. The fundamental elements of geology, such as rift formation and mountain ranges characterized by folding, can be readily understood as necessary surface adjustments to decompression-increased planetary volume. But this, to use a geological metaphor, is just the tip of the proverbial iceberg. There is more, lots more. Next we must consider the nature of Earth's interior as a consequence of condensing from within a giant gaseous protoplanet at high-pressures and high-temperatures.

Chapter 3 References

1.   Herndon, J.M., *Indivisible Earth: Consequences of Earth's Early Formation as a Jupiter-Like Gas Giant*, L. Margulis, Editor 2012, Thinker Media, Inc.

2.   Herndon, J.M., *Whole-Earth decompression dynamics*. Current Science, 2005. 89(10): p. 1937-1941.

3.   Herndon, J.M., *Energy for geodynamics: Mantle decompression thermal tsunami*. Current Science, 2006. 90(12): p. 1605-1606.

4.   Herndon, J.M., *Geodynamic Basis of Heat Transport in the Earth*. Current Science, 2011. 101(11): p. 1440-1450.

5.   Hilgenberg, O.C., *Vom wachsenden Erdball* 1933, Berlin: Giessmann and Bartsch. 56.

6.   Frank, L.A., *Atmospheric holes and small comets*. Rev. Geophys., 1993. 31(1): p. 1-28.

7.   Herndon, J.M., *Solar System processes underlying planetary formation, geodynamics, and the georeactor*. Earth, Moon, and Planets, 2006. 99(1): p. 53-99.

8.   Herndon, J.M., *Nuclear georeactor origin of oceanic basalt $^3He/^4He$, evidence, and implications*. Proceedings of the National Academy of Sciences USA, 2003. 100(6): p. 3047-3050.

9.   Herndon, J.M., *Nuclear georeactor generation of the earth's geomagnetic field*. Current Science, 2007. 93(11): p. 1485-1487.

10.   Mandea, M. and E. Dormy, *Asymmetric behavior of magnetic dip poles*. Earth Planet. Space, 2003. 55: p. 153-157.

11.   Herndon, J.M., *Origin of mountains and primary initiation of submarine canyons: The consequences of Earth's early formation as a Jupiter-like gas giant*. Current Science, 2012. 102(10): p. 1370-1372.

12.   Suess, E., *Das Antlitz der Erde* 1885, Prague and Vienna: F. Tempky.

13. Ryan, P.D. and J.F. Dewey, *Continental eclogites and the Wilson Cycle.* J. Geol. Soc. Lond., 1997. 154(3): p. 437-442.

14. Phillips, B.R. and H.-P. Bunge, *Supercontinent cycles disrupted by strong mantle plumes.* Geol., 2007. 35(9): p. 847-850.

15. Condie, K.C., *Breakup of a paleoproterozoic supercontinent.* Gond. Res., 2002. 5(1): p. 41-43.

16. Herndon, J.M., *Potentially significant source of error in magnetic paleolatitude determinations.* Current Science, 2011. 101(3): p. 277-278.

17. Brown, R., *On the formation of fjords, canons, and benches.* Proc. Roy. Geograph. Soc. Lond. , 1869. 13: p. 144-149.

18. Nesje, A. and I.M. Whillans, *Erosion of Sognefjord, Norway.* Geomorph., 1994. 9: p. 45.

19. Helland, A., *Die glaciale Buildung der Fjorde und Alpenseen in Norwegen.* Poggendorff's Ann. Physik Chemie, 1872. 146: p. 538-562.

20. Koechelin, R., *The formation of fjords.* J. Glaciol., 1947. 1: p. 66-68.

21. Porter, S.C., *Some geological implications of average Quaternary glacial conditions.* Quat. Res., 1989. 32: p. 245-261.

22. Shoemaker, E.M., *The formation of fjord thresholds.* J. Glaciol., 1986. 32: p. 65-71.

23. Troll, C., *Tiefenerosion, Seitenerosion und Akkumulation der Flusse im fluvioglazialen und periglazialen Bereich.* Petermanns Geogr. Mitt. Ergaenzunsh., 1957. 262: p. 213-226.

24. Gregory, J.W., *The Nature and Origin of Fjords*: (John Murray, 1913).

25. Harris, P.T. and T. Whiteway, *Global distribution of submarine canyons: Geomorphic differences between active and passive continental margins.* Marine Geol., 2011. 285: p. 69-86.

4. Nature of Earth's Interior

I admire the precision of your reasoning based on information available, and I congratulate you on the highly important result you have obtained. – *Inge Lehmann, inner core discoverer, to the author, 1979, Figure 4.3.*

Momentary streaks of light flashing across the night sky, called "shooting stars," are produced when meteors from outer space burn up as they come crashing to Earth. Occasionally, a meteor will survive the fiery trip through the atmosphere, land, and be recovered – then it is called a meteor*ite* – ite being a common suffix for rocks and minerals found on Earth. Meteorites provide crucial information about the nature of our planet and the Solar System. Although measurements of earthquake waves, augmented by studies of Earth's spin, can reveal the structure and physical states of matter within our planet, the chemical composition of that matter must be derived from investigations of meteorites.

In virtually every scientific endeavor, scientists, particularly those who work in laboratories, keep meticulous records so that later they can track their progress and repeat their procedures. In that way they know why and how they started, what materials, conditions, and equipment were employed and what they did to end up with the final results they obtained. Now, envision the formation of the Solar System about 4½ billion years ago occurring as if in an immense laboratory. There were plenty of events but yet no detailed records were kept, just the meteorites, the fragmental remains from the time when our Solar System was forming, clue-bearing bits and pieces. My approach to meteorite data is to arrange these seemingly independent clues into a logical sequence so that causal relationships become evident and reveal the nature of processes operant in that ancient time.

In 1850 Boisse [1] suggested that the composition of meteorites relates to the bulk composition of planet Earth. A boost of support for that idea came almost half a century later from the mind of the German seismologist, Emil Wiechert [2]. Wiechert lived during a time of extensive industrial iron-making in Europe and was acquainted with the process of molten iron metal settling beneath slag. Wiechert had seen meteorites in museums, some comprised entirely of iron metal, some a mixture of iron metal and stone, and some consisting solely of stone. (Figure 4.1) Realizing that the mean density of Earth, *i.e.*, its mass divided by its volume, as measured by Cavendish [3], was too great for the Earth to be composed entirely of rock, Wiechert suggested that the Earth has at its center a core of iron metal, like the metal of iron meteorites. Less than a decade elapsed before Wiechert's idea was confirmed by the 1906 discovery of the Earth's core, deduced from earthquake wave investigations by Richard Oldham [4]. Oldham's discovery was quickly followed by the 1909 discovery of Earth's very thin crust [5].

Treysa   Brahin   Gao-Guenie

Figure 4.1 The three types of meteorites as recognized circa end of the nineteenth century: iron, stony-iron, and stone, and their respective names. Photos courtesy of (left to right) Heinrich Stürzl, Steve Jurvetson, and Jon Taylor.

Over the next two decades, the dimension of Earth's core was precisely determined as was the fact that the core is liquid. But then, investigations of a large earthquake centered near New Zealand in 1929 produced a mystery that revealed a more complex view of Earth's interior.

When an earthquake wave passes from one material into another at an angle, its speed and direction are altered – much as light waves passing from air to water change angle, making a pencil in a half-full glass look like it's bent. In the case of the 1929 New Zealand earthquake, the change in direction upon entering and exiting the core should have resulted in a region on the Earth's surface, called the "shadow zone," where earthquake waves should not have been detected. But, mysteriously, earthquake waves were in fact detected within the shadow zone. In 1936, the Danish seismologist, Inge Lehmann, solved the mystery [6]. She reasoned that within the fluid core there exists a solid inner core about the size of the Moon. She showed that such a solid inner core would be capable of reflecting earthquake waves into the shadow zone. (Figure 4.2)

Figure 4.2 Inge Lehmann (1888-1993) and her discovery-diagram of Earth's inner core. For clarity, I re-traced the circles that bound the inner and fluid core, and the "shadow zone" region where earthquakes were claimed to be undetectable. Ray #5 is reflected into the shadow zone from the inner core which she envisioned existed.

Lehmann's discovery of the inner core immediately led to another puzzle: What is the composition of the inner core? The circa 1940 answer to that question by Francis Birch [7] and others initiated the modern and still current textbook version of Earth structure. But Birch and others made a mistake, a blunder subsequently compounded by decades of unquestioning acceptance by geophysicists that has led to much scientific confusion at significant expense to taxpayers.

About 80% of the meteorites that are observed falling to Earth are called "ordinary chondrites," thus-named because they are so common [8]. These are one of the three groups of meteorites whose elements were not appreciably separated from one another. Ordinary chondrites are composed of iron metal and silicate-rock, along with some iron sulfide. If an ordinary chondrite is heated to an elevated temperature, ~1300°C, the iron metal and iron sulfide meld forming a liquid alloy at temperatures at which the silicates are still solid. The denser liquid iron alloy would flow downward by gravity. Many scientists have assumed that the Earth as a whole is like an ordinary chondrite [7, 9, 10], which, superficially at least, would seem to account for the Earth having an iron alloy core surrounded by a silicate mantle. If so, then what is the composition of the inner core?

In ordinary chondrite meteorites, nickel occurs alloyed with iron metal. Elements heavier than nickel, even if taken together, could not make an object nearly as massive as the inner core. So, if nickel and iron are alloyed together in the core, and other elements are too low in abundance to form the inner core, then what is the composition of the inner core? This is the circa 1940 answer to that question: The inner core was therefore assumed to be solid iron metal that had begun to crystallize from the liquid iron core, like ice forming in a freezing glass of water. The difference between the inner core and fluid core was assumed to be a difference of crystalline state, not of chemical composition. This textbook interpretation of Earth's core has persisted essentially unchanged for decades, but cannot be correct: As we shall see, if Earth as a whole were to resemble an ordinary chondrite, it would have a smaller core than is observed, and that is not all. The geoscience community ignored the metal-rich, rare enstatite chondrites that account for only 2% of observed meteorite falls [8] and that are strikingly unlike ordinary chondrites in their oxidation state. Ignoring the rare and highly-reduced enstatite chondrites proved to be a mistake that led, not to advances, but to confusion. In science one should be extremely cautious about dismissing and the concomitant possibilities. What if the dismissed data were the correct data? How could one progress? Every follow-on "advance" based upon the non-dismissed, incorrect data would be incorrect. It would be like being trapped in a cul-de-sac with no place to go. This is what happened since 1940 in deep-Earth geophysics.

Nearly four decades after Birch's ideas were entrenched in the literature of geophysics, I investigated enstatite chondrites. Combined with two 1960s discoveries, my studies generated a fundamentally different view of Earth's inner core composition. What was the news? What were the two discoveries?

- **Elemental silicon occurs in the metallic iron of enstatite chondrites [11].**
- **Perryite, nickel-silicide, Ni_2Si, is present in many enstatite-chondrite meteorites [12, 13].**

I realized that in Earth's core, silicon in chemical combination with nickel would have settled by gravity to the center and, in principle, formed a mass virtually identical to the relative mass deduced for the inner core. When my nickel-silicide-inner-core concept was accepted for publication [14] (Appendix A), Inge Lehmann, the brilliant Dane who discovered Earth's solid inner core, commented favorably on the "highly important result." (Figure 4.3)

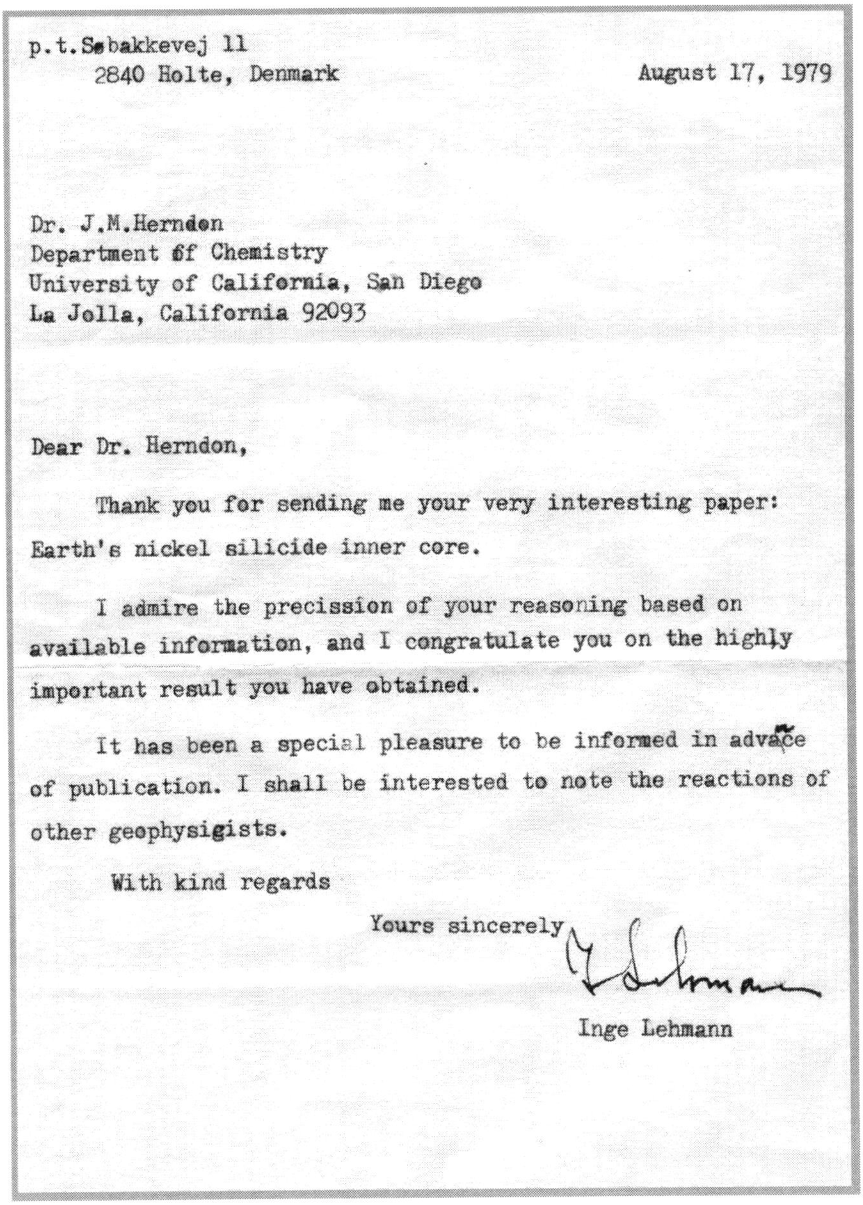

p.t.Søbakkevej 11
    2840 Holte, Denmark   August 17, 1979

Dr. J.M.Herndon
Department of Chemistry
University of California, San Diego
La Jolla, California 92093

Dear Dr. Herndon,

    Thank you for sending me your very interesting paper:
Earth's nickel silicide inner core.

    I admire the precission of your reasoning based on
available information, and I congratulate you on the highly
important result you have obtained.

    It has been a special pleasure to be informed in advace
of publication. I shall be interested to note the reactions of
other geophysigists.

    With kind regards

                  Yours sincerely,

                            Inge Lehmann

Figure 4.3 Scan of the congratulatory letter sent to me by Inge Lehmann.

At the time I no idea just how "highly important" the result would become, although I saw clearly how to show conclusively, if the Earth is like a chondrite meteorite as widely believed, it must be like an enstatite, not an ordinary, chondrite: The parts of the Earth must match the parts of an enstatite chondrite.

So, does it really matter that the Earth resembles an enstatite chondrite rather than an ordinary chondrite? Does it make any difference? The answer to each question is a resounding YES!

Science is very much a logical progression through time. Advances are frequently underpinned by ideas and understandings developed in the past, sometimes under circumstances which may no longer hold the same degree of validity. All too often, scientists, being distinctly human creatures of habit, plod optimistically along through time, eagerly looking toward the future, but rarely looking with question at circumstances from the past that have set them upon their present courses. Progressing along a logical path of discovery is rather like following a path through the wilderness. Occasionally, one comes to a juncture, the path splits, presenting a choice of scientific interpretations. Choose the correct logical interpretation and the way is clear for further progress; the wrong choice leads to confusion; a situation not unlike attempting to navigate a series of New York street addresses using a London city map. That is often the way of science. To make matters even more complicated, the correct path is sometimes invisible, obscured because some requisite discovery has not yet been made. Moreover, the logical progression of scientific discovery is often opposed by the darker elements of human nature, malevolent agendas as I described them in Chapter 1, and personal and institutional self-interest. This has been the case, I submit, with institutional impediments in our understanding of the composition and nature of Earth's deep interior – a quest I am interested in mostly for the sake of knowledge itself, but which also may have practical consequences.

Believing that the inside of Earth is like an ordinary chondrite meteorite, as many did and do (if not consciously then as a working assumption), led and leads to the incorrect conclusion that the silicate-rock mantle must be made of only one kind of rock below the crust and above the core. But, as early as the late 1930s, the New Zealand-born, Australian seismologist, Keith Bullen, had discovered that the silicate-rock mantle is not featureless, as it first had seemed [15, 16]. Bullen discovered that there is a seismic boundary or discontinuity at a depth of 660 km which separates the mantle into two major parts, upper and lower. Between that boundary and the core, the lower mantle, as it now called, comprises about 49% of the mass of the Earth and is without seismic irregularities. The upper mantle, lying above that boundary and comprising only 18% of the mass of the Earth, later was discovered to have several seismic boundaries, reminiscent of layers of veneer. So the question that faced geoscientists, who believe that the Earth is like an ordinary chondrite meteorite and that the mantle is of uniform chemical composition, is how can those seismically-indicated layers be explained?

Unless reflected, earthquake waves change speed and direction in a major way for just two reasons: They enter a zone at an angle that has either 1) a different crystal structure or 2) a different chemical composition. The seismic boundary in the mantle, discovered by Bullen, cried out for an explanation. Believing that the mantle has a uniform chemical composition, as Birch and others did, leads to the only explanation possible, namely, is that the seismic boundary that separates the mantle into two parts arises from changes in crystal structure caused by the pressure exerted by the weight of the rock above. Later, other seismic boundaries were discovered in the mantle and these too were afforded the same explanation. This is more of the decades-old geoscience textbook story, but, I discovered, it is not correct.

After a stimulating telephone conversation with Inge Lehmann in 1979, I progressed through the following logical exercise: If the inner core is in fact the compound nickel silicide, as I had suggested [14], then the Earth's core must be like the alloy portion of an enstatite chondrite. Then, if the Earth's core is in fact like the alloy portion of an enstatite chondrite, the Earth's core should be surrounded by a silicate-rock shell like the silicate-rock portion of an enstatite chondrite. But the enstatite chondrite type of silicate-rock is essentially devoid of FeO, which is iron combined with oxygen, unlike the silicate-rock of the upper part of the upper mantle, which has appreciable FeO [17]. This enstatite-chondrite-like silicate-rock shell, if it exists, thus should be bounded by a seismic discontinuity, the boundary where earthquake waves change speed and direction because of the different chemical compositions. So now here is a prediction which can be tested.

I tested it: Using the alloy to silicate-rock ratio of the Abee enstatite chondrite [18] and the mass of the Earth's core, by simple ratio proportion I calculated the mass of that enstatite-chondrite-like silicate-rock mantle shell. From tabulated mass distributions [19], I then found the radius of that predicted seismic boundary lays within about 1.2% of the radius of the major seismic discontinuity, occurring at a depth of about 660 km, the one which separates the lower mantle from the upper mantle. That logical exercise led me to discovery fundamental quantitative mass ratio relationships that connect the interior parts of the Earth with parts of the Abee enstatite chondrite [20-22].

Not only did I show that the inner 82% of Earth resembles an enstatite chondrite meteorite, but also that the Earth is not like an ordinary chondrite as had been long believed. This is how I did that: Metal-bearing chondrites generally consist of three components: nickel-iron alloy, iron sulfide, and silicates. If one were to heat either an ordinary chondrite or an enstatite chondrite to a sufficiently high temperature, the iron metal and iron sulfide would combine and liquefy, settling by gravity beneath the less-dense silicate part, just as steel settles beneath slag on a steel-hearth. The Earth is like a spherical steel-hearth, its entire core or alloy part comprising 32.5% of the planet's mass. Some enstatite chondrites have a sufficiently high percentage of iron-alloy to make such a massive core; but no ordinary chondrites do [20, 21, 23]. (Figure 4.4)

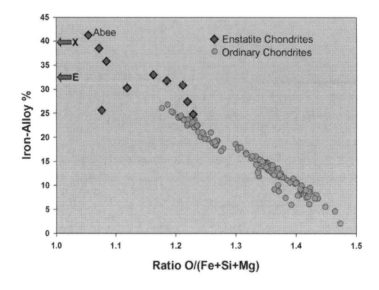

Figure 4.4. Evidence that Earth is like an enstatite chondrite. The percent alloy (iron metal plus iron sulfide) of 157 ordinary (circles)- and 9 enstatite-chondrite-meteorites (diamonds) plotted against oxygen content. The core percent of the whole-Earth, 'arrow E', and of (core-plus-lower mantle), 'arrow X', shows that Earth is in the main like an Abee-type enstatite chondrite and not at all like an ordinary chondrite.

Dahm [24] and Bullen [16] first discussed the possibility of some seismic irregularity at the boundary between the core and the lower mantle. Subsequent investigations confirmed the existence of "islands" of matter at the boundary of the core [25, 26]. This seismic irregularity or "roughness" has been ascribed to a pressure-induced change in crystal structure [27, 28]. By contrast, I showed that "islands" of matter at the core-mantle boundary are readily understandable in a logical and consequentially related way as low-density, high-temperature precipitates from the Earth's enstatite-chondrite-like core [22, 23, 29].

Figure 4.5 presents the relative abundances of major and minor elements of the Abee enstatite chondrite, as ratios relative to iron, showing their distribution between silicate and alloy portions. (With ratios one can, for example, compare the parts of a meteorite with the parts of the Earth.) In the more oxygen-rich ordinary chondrites, all of the silicon, calcium and magnesium occur as oxides in the silicate portion. But, matter like that of the Abee enstatite chondrite formed under conditions that severely limited available oxygen. Consequently, the major alloy of enstatite chondrites contains some calcium, magnesium and silicon, in addition to iron, nickel, and sulfur.

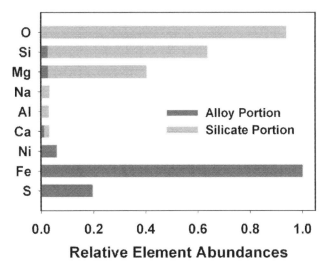

Figure 4.5 Relative abundances of the major and minor elements in the Abee enstatite chondrite, normalized to iron, showing their relative amounts in the alloy and silicate portions. Note that calcium (Ca), magnesium (Mg), and silicon (Si), occur in part in the alloy portion, whereas in ordinary chondrites they occur entirely in the silicate part.

Imagine Earth's core, composed of enstatite-chondrite-like alloy, so hot that all of those elements are dissolved into an iron-based alloy. What happens as this alloy begins to cool? Some elements, such as sulfur, are compatible with being dissolved in molten iron; they will tend to stay dissolved to a relatively low temperature. But, some other elements are quite *incompatible*, particularly oxygen-loving elements. Incompatible elements, like calcium and magnesium, in a cooling iron alloy solution will seek a way to come out of solution, and they find it by combining with sulfur. Both calcium sulfide (CaS) and magnesium sulfide (MgS) form solids at temperatures that are well above the melting point of pure iron (1538°C).

Both are less dense than iron and will float atop the molten iron. I have suggested that calcium sulfide (CaS) and magnesium sulfide (MgS) precipitated from Earth's core and floated to its top, causing the seismic "roughness," *i.e.*, the "islands," at the core-mantle boundary [22, 23, 29]. The observed abundances are appropriate and, moreover, there is an industrial process which is quite similar [30-32]. Sulfur impurity can weaken steel. So, to remove sulfur from high-quality steel, magnesium or calcium is injected into the molten iron to combine with the sulfur at a high temperature and float to the surface.

Figure 4.6 shows that the parts of the interior of the Earth match quite precisely the parts of the primitive, Abee enstatite chondrite. This is powerful evidence from which one may conclude:

- **In the main, Earth resembles an enstatite chondrite, *not* an ordinary chondrite.**
- **The seismic boundaries below a depth of 660 km are caused by differences in chemical compositions, not simply a difference in crystal structure.**
- **The deep interior of Earth has a highly-reduced state of oxidation like that of the primitive enstatite chondrites, *i.e.*, its minerals contain less oxygen than, for instance, those of an ordinary chondrite.**

Figure 4.7 is a schematic representation of the chemical compositions of the various major parts of the Earth. The compositions of the lower mantle and core can be deduced from the identities shown in Figure 4.6. The compositions of the layers of the upper mantle, on the other hand, are yet unknown, but may include components from ordinary and carbonaceous chondrite matter, as indicated, for example, by the oxidized iron found in rocks brought from the upper mantle by volcanoes.

Fundamental Earth Ratio	Earth Ratio Value	Abee Ratio Value
lower mantle mass to total core mass	1.49	1.43
inner core mass to total core mass	0.052	theoretical 0.052 if Ni_3Si 0.057 if Ni_2Si
inner core mass to lower mantle + total core mass	0.021	0.021
D" mass to total core mass	0.09	0.11
ULVZ of D" CaS mass to total core mass	0.012	0.012

Figure 4.6 Fundamental mass ratio comparison between the endo-Earth (lower mantle plus core) and the Abee enstatite chondrite. D" is the "seismically rough" region between the fluid core and lower mantle. ULVZ is the "Ultra Low Velocity Zone" of D". Average measurements of the Abee, Indarch, and Adhi-Kot enstatite chondrites are used for D". Data from [18, 41, 42]. For more information, see [22].

In 1944, on the basis of thermodynamic considerations, Eucken [33] suggested core-formation in the Earth as a consequence of successive condensation from solar matter, on the basis of volatility, from the central region of a hot, gaseous protoplanet with molten iron metal first condensing and falling in to the center. But what was missing in Eucken's portrayal was a way to tie those calculations directly to the matter of Earth.

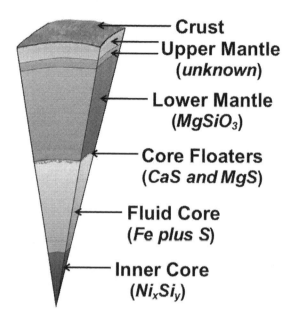

Figure 4.7 Chemical compositions of the major parts of the endo-Earth. The nature of the compositions of parts within the upper mantle are presently unknown.

I have provided that way by showing, via the identities presented in Figure 4.6, that the matter comprising the interior of Earth is essentially identical to the matter of primitive enstatite chondrite meteorites, and by showing from thermodynamic considerations that the state of oxidation of such matter results from condensation [34-36] under the same conditions as described by Eucken [33] for Earth, namely, condensation from solar matter at high-temperatures and high-pressures with concomitant in-falling and accumulating at the center, that is to say "raining-out." I generalized that condensation concept to other massive-core planets [37] and showed there is a commonality in planetary compositions and processes. The evolutionary biologist, Lynn Margulis taught the importance of considering Earth as a whole, not as separate pieces spread among different disciplines. I adopt that approach in my work and in this book, not only with Earth, but in the broader framework of our Solar System as discussed in Chapter 8.

Indeed, evidence of planetary formation by condensing and raining-out from solar matter at high-temperatures and high-pressures may already have been discovered through recently observed "pits" on Mercury's surface.

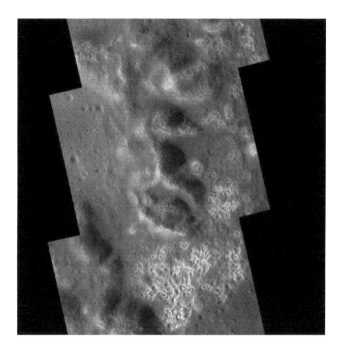

Figure 4.8 NASA MESSENGER image, taken with the Narrow Angle Camera, shows an area of hollows on the floor of Raditladi basin on Mercury. Surface hollows were first discovered on Mercury during MESSENGER's orbital mission and have not been seen on the Moon or on any other rocky planetary bodies. These bright, shallow depressions appear to have been formed by disgorged volatile material(s) from within the planet.

Many of the images from the NASA's Project MESSENGER spacecraft, shown in Figures 4.8 and 4.9, reveal [38]:

> ... an unusual landform on Mercury, characterized by irregular shaped, shallow, rimless depressions, commonly in clusters and in association with high-reflectance material ... and suggest that it indicates recent volatile-related activity.

But the authors were unable to describe a scientific basis for the source of those volatiles, stating:

> ... Mercury's interior contains higher abundances of volatile elements than are predicted by several planetary formation models for the innermost planet.

Furthermore, the authors were unable to suggest identification of the "high-reflectance material."

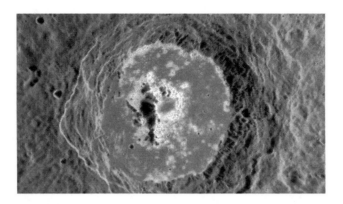

Figure 4.9 NASA MESSENGER image of a complex crater exhibits many hollows along its floor and central peak complex. The hollows have a very high albedo, which makes this crater stand out prominently.

I immediately felt I understood those observations and described their basis in a scientific communication entitled "Hydrogen Geysers: Explanation for Observed Evidence of Geologically Recent Volatile-related Activity on Mercury's Surface" [39]. I provided the background on the my concept of Mercury's formation by raining out from within a giant gaseous protoplanet at hydrogen pressures in excess of one atmosphere and the subsequent stripping away of primordial gases. I showed that primordial condensation from an atmosphere of solar composition, at pressures above about one atmosphere, leads to iron metal condensing as a liquid. (Figure 4.10) It is well known in metallurgy that molten iron is capable of dissolving copious amounts of hydrogen. My calculations indicate that one or more Mercury-size volumes of hydrogen at Standard Temperature and Pressure (STP) could potentially dissolve in the iron that rained-out to form Mercury's core, which is later released as the core solidifies. The release of hydrogen escaping at the surface, I posited, is responsible for the formation of said "*unusual landform on Mercury,*" sometimes referred to as pits, and for the formation of the associated "*high-reflectance* material," bright spots, which I suggested is iron metal reduced from an iron compound, probably iron sulfide, by the escaping hydrogen.

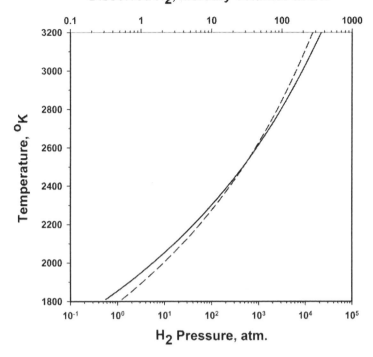

Dissolved H$_2$, Mercury-volumes at STP

Figure 4.10 The solid curve in this figure shows the temperatures and hydrogen pressures in a cooling atmosphere of solar composition at which liquid iron ideally will begin to condense. The dashed curve shows the substantial volume of hydrogen ideally that could dissolve in Mercury's iron in that medium at the temperatures and pressures indicated by the solid curve. Dissolved hydrogen is expressed as volume at Standard Temperature and Pressure (STP) and expressed as multiples of planet Mercury's volume.

The amount of dissolved hydrogen released during Mercury's core solidification is certainly sufficient to account for the *"unusual landform"* on Mercury's surface and may also be involved in bringing up iron sulfide, which is abundant on the planet's surface. The hydrogen may chemically combine with some sulfur thus changing iron sulfide into iron metal. The shiny iron metal thus accounts for the associated *"high-reflectance material,"* bright spots. So, here is a test: Proving that the *"high-reflectance material"* is indeed low-nickel metallic iron would provide strong evidence that the exhausted gas is hydrogen, and that protoplanetary condensation took place at pressures at or above about one atmosphere [34, 35, 37]. In principle that might be done, even remotely, by measuring the spectrum of light that reflects from the shiny surface and comparing that spectrum with a known laboratory reflectance spectrum from iron metal.

It is hard to imagine any two planetary theories that could be any more different than:

- **The long-popular "standard model" planetesimal theory with its ordinary-chondrite-like Earth composition and associated plate tectonics surface behavior;**
- **The protoplanetary condensation theory I set forth with its enstatite-chondrite-like Earth composition and Whole-Earth Decompression Dynamics surface behavior**

The major difference is that the latter above, WEDD, is derived logically from fundamental quantitative relationships, such as shown in Figure 4.6, as well as Earth's early existence as a Jupiter-like gas giant. By contrast, the former, long-popular version, is comprised of arbitrary bits and pieces, glued together with *ad hoc* assumptions.

Take, for example, the nature of the energy source that geomagnetic evidence indicates exists at or near the center of Earth. A magnetic field is something that is invisible, something which we are unable to see or feel. We can envision the shape of a magnetic field by the way iron-filings are oriented by a hand-magnet, as illustrated in Figure 4.11. Our entire planet is surrounded and, indeed, protected from the solar wind by the Earth's magnetic field, also known as the geomagnetic field. The geomagnetic field directs magnetized needles of compasses to point in a constant direction as the compass is rotated. Certain migratory birds and butterflies, as recently discovered, use the geomagnetic field to obtain directions, although there is some mystery as to precisely how that works. Similarly, in the ocean, whales and dolphins use their built-in magnetic compasses to obtain directions from the Earth's magnetic field. Even some bacteria, called magnetotactic bacteria, have built-in nano-size magnetic compasses, made of magnetite, that allow them to navigate using the geomagnetic field.

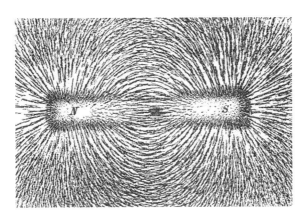

Figure 4.11 Iron filings on a piece of cardboard with a hand-magnet below will align themselves with its magnetic field and thus provide a visual representation of the invisible magnetic field.

The geomagnetic field was deduced by one, William Gilbert. Gilbert was a physician of prominence who eventually became the personal physician to Queen Elizabeth I. But his lasting achievement was not in medicine, but resulted from discoveries he made investigating magnetism. He collected and published magnetic data that mariners had collected on their distant voyages. He also had a sphere fabricated from lodestone, the mineral magnetite. The loadstone was thought to have become magnetized by a near-by lightning strike. Gilbert passed a magnetic compass about the loadstone-sphere. He found that the directions of the compass about the lodestone sphere were quite similar to observations made by navigators throughout the world. Those observations convinced Gilbert that the origin of geomagnetic field was neither magic nor extraterrestrial, as some thought at the time, but rather part of the Earth. Publishing his work in Latin, as was common practice at the time. Gilbert named his book *De Magnete, Magneticisque Corporibus, et de Magno Magnete Tellure (On the Magnet and Magnetic Bodies, and on the Great Magnet the Earth)*.

In 1828, at the suggestion of Alexander von Humboldt, Carl Friedrich Gauss, the greatest German mathematician of his time, began to apply his formidable talents toward understanding the geomagnetic field. His great achievement was in showing that the magnetic measurements made at Earth's surface could be described mathematically by the equation for a dipole (bar magnet) located at or near the center of the Earth. (Figure 4.12)

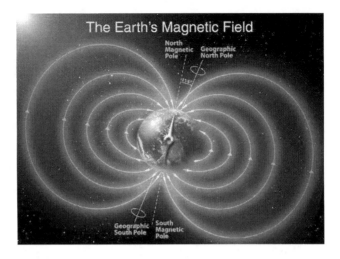

Figure 4.12 Schematic representation of Earth's magnetic field as if formed by a magnetic at its center. The interior, however, is too hot for a permanent magnet to exist. Image courtesy of Peter Reid.

The mysterious geomagnetic field, once thought to originate beyond Earth or to be caused by a polar mountain of magnetite, originates at the center of Earth. Now, this became an even greater mystery. Why? First, a permanent magnet, like a bar magnet, cannot remain magnetic at temperatures above about 700°C. Second, a permanent magnet would soon lose its magnetism because interactions of the geomagnetic field with matter of the Earth and with charged particles of the solar wind would drain energy from the magnetic field. So, at or near the center of Earth there must exist a mechanism for continuously producing the geomagnetic field: some power source must work to continuously revitalize it.

There are three elements, uranium, thorium, and potassium that are radioactive and have long been considered possible sources of energy for the geomagnetic field. In the long-popular planetesimal theory with its ordinary-chondrite-like Earth composition and associated plate tectonics surface behavior, uranium and thorium typically were discounted. In ordinary chondrites those three elements universally occur in the silicates, not in the iron-alloy portion. From time to time suggestions have been advanced that deep-Earth pressures might allow potassium to reside in the Earth's core, but no natural mechanism for its concentration has been proposed.

So what then has been proposed for the deep-Earth energy source that powers the geomagnetic field? The assumed energy source is based upon a series of assumptions, the main ones of which are 1) the inner core is iron metal and 2), that the inner core is cooling and growing (by the crystallization of iron), which is imagined to provide the necessary energy. There is no evidence for either of these long-popular but still *ad hoc* mechanisms.

On the other hand in our new geoscience paradigm with its enstatite-chondrite-like Earth composition a powerful energy source at or near the center of Earth is a natural consequence of our original understanding, obviating the need to make improvised assumptions.

Let us look at how the geomagnetic field is a natural consequence of Earth's enstatite-chondrite-like composition that would be expected from protoplanetary condensation and formation. Murrell and Burnett [40] discovered that virtually all of the uranium and thorium in the Abee enstatite chondrite occurs in minerals corresponding to the alloy portion of that meteorite. By the relationships shown in Figure 4.10, that means that as much as 82% of the Earth's inventory of uranium and thorium still exists trapped within our planet's core. And that's not all. As described in the next chapter, there is evidence that uranium settling to Earth's center functions as a natural georeactor, nuclear fission reactor, providing not only the required energy source, but a way for generating the Earth's geomagnetic field.

Chapter 4 References

1. Boisse, A.A.M., *Recherches sur l'histoire, la nature et l'origine des aérolilbes* Rodes (Ratery), 1850. 12 Bog. gr. 8.

2. Wiechert, E., *Ueber die Massenverteilung im Inneren der Erde.* Nachr. K. Ges. Wiss. Goettingen, Math.-Kl., 1897: p. 221-243.

3. Cavendish, H., *Experiments to determine the density of Earth.* Phil. Trans. Roy. Soc. Lond., 1798. 88: p. 469-479.

4. Oldham, R.D., *The constitution of the interior of the earth as revealed by earthquakes.* Q. T. Geol. Soc. Lond., 1906. 62: p. 456-476.

5. Mohorovicic, A., Jb. Met. Obs. Zagreb, 1909. 9: p. 1-63.

6. Lehmann, I., *P'.* Publ. Int. Geod. Geophys. Union, Assoc. Seismol., Ser. A, Trav. Sci., 1936. 14: p. 87-115.

7. Birch, F., *The transformation of iron at high pressures, and the problem of the earth's magnetism.* Am. J. Sci., 1940. 238: p. 192-211.

8.   Grady, M.M., *Catalogue of Meteorites*, 2000, Cambridge University Press: Cambridge, England. p. 689.

9.   Birch, F., *Differentiation of the mantle*. Bull. Geol. Soc. Am., 1958. 69: p. 483-486.

10.   Bullen, K.E., *Compressibility-pressure hypothesis and the Earth's interior*. Mon. Not. R. Astron. Soc., Geophys. Suppl., 1949. 5: p. 355-398.

11.   Ringwood, A.E., *Silicon in the metal of enstatite chondrites and some geochemical implications*. Geochim. Cosmochim. Acta, 1961. 25: p. 1-13.

12.   Ramdohr, P., *Einiges ueber Opakerze im Achondriten und Enstatitachondriten*. Abh. D. Akad. Wiss. Ber., Kl. Chem., Geol., Biol., 1964. 5: p. 1-20.

13.   Reed, S.J.B., *Perryite in the Kota-Kota and South Oman enstatite chondrites*. Mineral Mag., 1968. 36: p. 850-854.

14.   Herndon, J.M., *The nickel silicide inner core of the Earth*. Proc. R. Soc. Lond, 1979. A368: p. 495-500.

15.   Bullen, K.E., *Note on the density and pressure inside the Earth*. Trans. Roy. Soc. New Zealand, 1938. 67: p. 122.

16.   Bullen, K.E., *A hypothesis on compressibility at pressures on the order of a million atmospheres*. Nature, 1946. 157: p. 405.

17.   Harte, B., *Kimberlite nodules, upper mantle petrology, and geotherms*. Phil. Trans. R. Soc. Lond., 1978. A 288: p. 487-500.

18.   Keil, K., *Mineralogical and chemical relationships among enstatite chondrites*. J. Geophys. Res., 1968. 73(22): p. 6945-6976.

19.   Dziewonski, A.M. and F. Gilbert, *Observations of normal modes from 84 recordings of the Alaskan earthquake of 1964 March 28*. Geophys. Jl. R. Astr. Soc., 1972. 72: p. 393-446.

20.   Herndon, J.M., *The chemical composition of the interior shells of the Earth*. Proc. R. Soc. Lond, 1980. A372: p. 149-154.

21. Herndon, J.M., *Composition of the deep interior of the earth: divergent geophysical development with fundamentally different geophysical implications.* Phys. Earth Plan. Inter, 1998. 105: p. 1-4.

22. Herndon, J.M., *Geodynamic Basis of Heat Transport in the Earth.* Current Science, 2011. 101(11): p. 1440-1450.

23. Herndon, J.M., *Scientific basis of knowledge on Earth's composition.* Current Science, 2005. 88(7): p. 1034-1037.

24. Dahm, C.G., *A Study of Dilatational Wave Velocity in Earth as a Function of Depth,* 1934, St. Louis University: St. Louis, MO.

25. Lay, T. and D.V. Helmberger, *The shear wave velocity gradient at the base of the mantle.* J. Geophys. Res., 1983. 88: p. 8160-8170.

26. Vidale, J.E. and H.M. Benz, *Seismological mapping of the fine structure near the base of the Earth's mantle.* Nature, 1993. 361: p. 529-532.

27. Mao, W.L., et al., *Ferromagnesian postperovskite silicates in the D″ layer of the Earth.* Proc. Nat. Acad. Sci. USA, 2004. 101(45): p. 15867-15869.

28. Murakami, M., et al., *Post-perovskite phase transition in $MgSiO_3$.* Sci., 2004. 304: p. 855-858.

29. Herndon, J.M., *Feasibility of a nuclear fission reactor at the center of the Earth as the energy source for the geomagnetic field.* J. Geomag. Geoelectr., 1993. 45: p. 423-437.

30. Foster, E., et al., *Deoxidation and desulphurization by blowing of calcium compounds into molten steel and its effects on the mechanical properties of heavy plates.* Stahl u. Eisen, 1974. 94: p. 474.

31. Inoue, R. and H. Suito, *Calcium desulfurization equilibrium in liquid iron.* Steel Res., 1994. 65(10): p. 403-409.

32. Ribound, P. and M. Olette. *Desulfurization by alkaline-earth elements and compounds.* in *Physical Chemistry and Steelmaking.* 1978. Versailles, France.

33. Eucken, A., *Physikalisch-chemische Betrachtungen ueber die frueheste Entwicklungsgeschichte der Erde*. Nachr. Akad. Wiss. Goettingen, Math.-Kl., 1944: p. 1-25.

34. Herndon, J.M., *Solar System processes underlying planetary formation, geodynamics, and the georeactor*. Earth, Moon, and Planets, 2006. 99(1): p. 53-99.

35. Herndon, J.M., *Nature of planetary matter and magnetic field generation in the solar system*. Current Science, 2009. 96(8): p. 1033-1039.

36. Herndon, J.M. and H.E. Suess, *Can enstatite meteorites form from a nebula of solar composition?* Geochim. Cosmochim. Acta, 1976. 40: p. 395-399.

37. Herndon, J.M., *New indivisible planetary science paradigm*. Current Science, 2013. 105(4): p. 450-460.

38. Blewett, D.T., et al., *Hollows on Mercury: MESSENGER Evidence for Geologically Recent Volatile-Related Activity*. Science, 2011. 333: p. 1859-1859.

39. Herndon, J.M., *Hydrogen geysers: Explanation for observed evidence of geologically recent volatile-related activity on Mercury's surface*. Current Science, 2012. 103(4): p. 361-361.

40. Murrell, M.T. and D.S. Burnett, *Actinide microdistributions in the enstatite meteorites*. Geochim. Cosmochim. Acta, 1982. 46: p. 2453-2460.

41. Dziewonski, A.M. and D.A. Anderson, *Preliminary reference Earth model*. Phys. Earth Planet. Inter., 1981. 25: p. 297-356.

42. Kennet, B.L.N., E.R. Engdahl, and R. Buland, *Constraints on seismic velocities in the earth from travel times* Geophys. J. Int., 1995. 122: p. 108-124.

5. Georeactor Heat Production

Herndon's idea about georeactor located at the center of the Earth, if validated, will open a new era in planetary physics. – *G. Domogatski, V. Kopeikin, L. Mikaelyan, and V. Sinev [1].*

The existence of a geocentric nuclear fission reactor is one consequence of Earth formation from within a giant gaseous protoplanet by condensing at high-pressures and high-temperatures, in-falling to the center, and accumulating there. In 1993, I applied Fermi's nuclear reactor theory to demonstrate the feasibility of the existence of a georeactor at Earth's center. Since then I have continued to explore the consequences of this heterodox proposition. In 2014, I published, in *Current Science*, a comprehensive review article on the subject [2]. (Appendix B is a copy of that review article.)

Before 1896 scientists believed without question that the chemical elements were forever permanent, medieval alchemical dreams of changing base metals to gold notwithstanding. But in 1896 a serendipitous discovery was made by French Nobel laureate physicist and co-discover (with Marie Curie) of radioactivity, Henri Becquerel, investigating a process, now called phosphorescence. In phosphorescence certain substances absorb energy from sunlight and then re-radiate it at a later time. Becquerel would expose a chemical to sunlight then place it against a photographic plate in a light-proof container. After a time, he would develop the photographic plate to observe whether delayed re-radiation of light from the chemical had darkened the plate. On one occasion he realized that the chemical, a salt of uranium, had been exposed to sunlight on a cloudy day, but instead of discarding the photographic plate, he proceeded to develop it and was amazed by the amount of darkening that had occurred. In follow-up experiments he showed that uranium was emitting energy from within: Becquerel had discovered radioactivity, the spontaneous change of one element into another with concomitant release of energy [3].

During the first four decades of the 20th century there was a revolution in physics. New discoveries and new insights broadened and deepened our understanding about the nature of matter and radiation, including radioactivity. For background I will briefly describe in simplified form just a few consequences of these many and diverse scientific investigations.

Think of an atom as being comprised of an electrically positive, massive nucleus surrounded at some distance away by shells of orbiting, negatively charged electrons. The nucleus is comprised of positively charged protons held tightly together in some way by their association with neutrons, which have no electric charge. The number of protons is the same as the number of electrons, which determines the chemical behavior of the electrically neutral atom. Instead of having a name, a chemical element could be called by the number of protons in its nucleus. For example, hydrogen could be called #1; oxygen, #8; uranium #92, and so forth. The number of neutrons in the atomic nucleus tends to be a slight increment greater than the proton number, with that slight increment increasing as the proton number increases. (Figure 5.1)

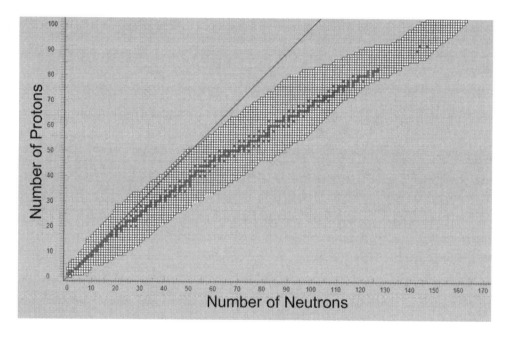

Figure 5.1 Plot of proton number versus neutron number for all possible isotopes of all of all chemical elements. The filled squares represent the compositions of stable isotopes, and include the long-lived radioactive isotopes of thorium and uranium. The open squares represent the compositions of possible radioactive isotopes, most of which have very short half-lives. Radioactive isotopes eventually decay to become stable isotopes. The straight line shows equal numbers of protons and neutrons. Note that as the proton number increases, the number of neutrons increases by a greater proportion. Image courtesy of Joe Magill and Nucleonica.

Some elements are found in nature with a single, unique number of neutrons, for example, fluorine (F) with 9 protons and 10 neutrons or gold (Au) with 79 protons and 116 neutrons. Other elements are found with more than one possible number of neutrons, for example, chlorine (Cl) with 17 protons and either 18 or 20 neutrons; these various forms of the same element are referred to as "isotopes." In other words, fluorine has one isotope; chlorine has two isotopes. Expressed in a shorthand way of writing, these are, for fluorine, ^{10}F, and, for chlorine, ^{18}Cl and ^{20}Cl. Tin (Sn) has 10 isotopes, the most of any stable element.

Not all isotopes are stable, though. Some are radioactive, meaning that they decay naturally to become different elements, changing their proton numbers, but by no more than two in one decay step. And when they decay they release energy. The average radioactive decay rate, unique and specific for each radioactive isotope, is expressed by its half-life, which is to say the time required for the isotope to diminish by 50%. Because of their abundances and relatively long half-lives, the following isotopes of uranium, thorium, and potassium have long been thought to be the most important radioactive heat sources within the Earth: ^{235}U, ^{238}U, ^{232}Th, and ^{40}K.

As early investigations showed, nuclear reactions can be artificially induced by bombarding a target nucleus with neutrons. This may cause the target element to become a different element, changing its proton number, but by no more than two. Then, in 1938 Otto Hahn and Fritz Strassmann bombarded uranium with neutrons and chemically detected barium in the irradiated target. Barium with proton number 56 is roughly half the proton number of uranium, 92. This was no ordinary nuclear reaction; Hahn and Strassmann had discovered nuclear fission, the splitting of the atom [4]. It was soon realized that nuclear fission of uranium releases an enormous amount of energy plus it liberates a few neutrons, which makes possible the nuclear fission chain reaction.

Imagine: A neutron splits a uranium nucleus, which liberates a couple of neutrons, one of which splits another uranium nucleus, which liberates a couple of neutrons, one of which splits another uranium nucleus, which liberates a couple of neutrons, one of which splits yet another uranium nucleus ... and so forth. (Figure 5.2) This is a nuclear fission chain reaction. If the chain reaction occurs uncontrollably and almost instantly, this is an atomic bomb. If the chain reaction is regulated and heat is removed as soon as produced, this is a nuclear fission reactor.

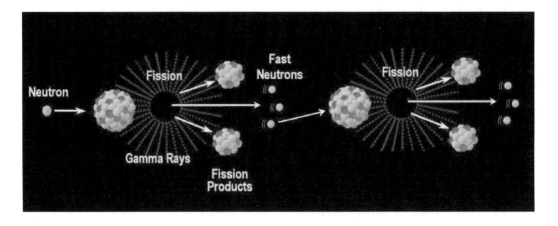

Figure 5.2 Schematic representation of the nuclear fission chain reaction. Image courtesy of Jaroslav Franta.

Nuclear reactor – the mention of those two words might bring to mind names like Chernobyl or Three Mile Island or perhaps conjure images of complex mega-machines whose control rooms have more instrumentation than the cockpit of a 747. But this complex instrumentation belies the simplicity at the basis of the operation which runs a nuclear reactor. (Figure 5.3)

Figure 5.3 Kozloduy Nuclear Power Plant - Control Room of Units 1 and 2, shut down in 2003. Photo courtesy of Yuvko Lambrev.

At its heart a nuclear reactor is simply an accumulation of very heavy atoms, such as uranium, whose nuclei will split when hit with a neutron. When hit with a neutron, the very-heavy atomic nucleus breaks into two parts, liberating energy plus a few more neutrons. These just-freed neutrons can strike another nucleus, causing it to fission; it can happen again and again and again, and on and on as a chain reaction. But if too many neutrons come out of play, by escaping or by being absorbed, a chain reaction cannot be sustained.

Enrico Fermi formulated nuclear reactor theory [5] and in 1942 designed and built the world's first man-made nuclear reactor. In 1956, using Fermi's nuclear reactor theory, Paul Kazuo Kuroda, showed that nuclear fission chain reactions could have occurred in uranium deposits two billion years ago and earlier [6, 7]. At the time of its proposal Kuroda's idea was very unpopular; indeed, Kuroda told me that the only way he managed to get it published was because, at the time he proposed it, the *Journal of Chemical Physics* would publish short papers without review.

Fast forward 16 years to 1972 when I was a graduate student. One day Marvin W. Rowe, my thesis advisor, rushed into the lab to tell me that his former thesis advisor, Paul K. Kuroda, had just learned that French scientists had discovered the intact remains of a natural nuclear reactor in a uranium mine at Oklo in the Republic of Gabon in Western Africa. (Figure 5.4)

Figure 5.4 Seam of uranium ore in an Oklo natural nuclear reactor zone. Photo courtesy of Francoise Gauthier-Lafaye.

The reactor had functioned two billion years ago just as Kuroda had predicted. Later, other fossil reactors were discovered in the region. I remember thinking at the time that the discovery must have huge implications. Indeed, it does. But at the time there were just too many pieces missing from the puzzle to progress further – it was like looking out into a very, very dense fog. Over the next two decades, however, without consciously planning to, I realized I had started to fill in some of the missing pieces.

Obviously, the first thing to realize is that Kuroda's prediction of a natural, non-manmade nuclear reactor was vindicated. Then, in 1980, I was able to show the likelihood that the inner 82% of the Earth resembles a primitive enstatite chondrite, such as the Abee meteorite. Because the deep interior of Earth, like the Abee meteorite, is highly-reduced, I wondered whether uranium and/or thorium might reside in Abee's alloy which corresponds to Earth's core. I submitted a research proposal to the U. S. National Science Foundation to investigate that question experimentally, but my proposal was declined, however, by anonymous reviewers who believed Earth is like an ordinary chondrite. So far, so bad for me and my crazy views. Then, good fortune struck. In 1982, Murrell and Burnett published measurements showing that most, if not all of the uranium and thorium in the Abee meteorite resides in the alloy portion [8].

Deep within the Earth, density depends only on atomic number and atomic mass. Uranium, precipitated from the fluid core, being the densest substance, would tend to settle by gravity to Earth's center. Here was a natural mechanism for concentrating the uranium! I applied Fermi's nuclear reactor theory to Earth-core uranium, much as Kuroda had done for veins of surface uranium [6, 7]. I then submitted my manuscript, entitled "**Feasibility of a Nuclear Fission Reactor at the Center of the Earth as the Energy Source for the Geomagnetic Field**," to the Japanese *Journal of Geomagnetism and Geoelectricity*. Never once in the review process did I encounter pejorative or caustic comments, or the negative tone and handwaving dismissals that I had unfortunately come to expect among American editors and referees. Instead, the editor and the referees made positive suggestions for improving the manuscript. One referee in particular insisted that I provide more detailed descriptions and additional calculations, which I did. The manuscript was published in 1993 [9]. Only after his death in 2001 did I learn that Paul Kazuo Kuroda was that insistent referee.

I continued to make advances in understanding the structure and behavior of the geocentric (or "Terracentric" as I called it in the *Current*

Science review article) nuclear fission reactor, or "georeactor" as it came to be known for short. I published one scientific article on the subject in 1994 in the *Proceedings of the Royal Society of London* [10] and another in 1996 in the *Proceedings of the National Academy of Sciences, USA* [11], two of the leading peer-reviewed scientific journals in the world. (Figure 5.5 is a schematic representation of the Earth's interior showing the georeactor.)

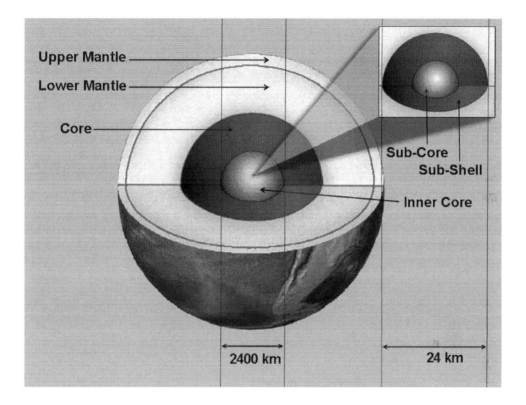

Figure 5.5 Schematic cut-away of Earth's interior and (inset) nuclear fission georeactor at Earth's center. The georeactor at the center is one ten-millionth the mass of Earth's fluid core. The georeactor's radioactive-waste sub-shell, I posit, is a liquid or slurry and is situated between the nuclear fission heat source and inner-core heat sink, assuring stable convection, which is necessary for sustained geomagnetic field production by convection-driven dynamo action in the georeactor sub-shell.

For over thirty years, engineers and scientists at Oak Ridge National Laboratory have developed software to conduct computer simulations of the operation of different types of nuclear reactors. Their software has been validated by comparing calculation results with analyses of spent nuclear reactor fuel rods. A major advance was made in 2001 when Daniel F.

Hollenbach agreed to modify the software, making two changes so as to accommodate the georeactor. One change allowed operating-times to extend to the age of the Earth and beyond. The other change made possible the removal of fission products. Why the latter change?

Deep inside planets and large moons, density depends only on atomic number and atomic mass. When a uranium nucleus splits, it produces two fission fragments, each having about half the atomic number and half of the atomic mass of the parent uranium. The fission fragments are thus less dense than uranium and migrate outward, forming a nuclear waste sub-shell surrounding the uranium sub-core.

The computer simulations made at Oak Ridge National Laboratory verified my previous calculations and expectations: The georeactor is indeed capable of functioning as a breeder-reactor (one which produces additional fissionable fuel from non-fissionable nuclides) over the entire age of the Earth at a power level within the estimated range for geomagnetic field production. Moreover, the Oak Ridge calculations provided data on the fission products that are unobtainable from calculations based upon Fermi's nuclear reactor theory. Helium fission product data led to the first substantial evidence for georeactor existence.

Helium is a mixture of two components (isotopes), helium of mass 3, ^{3}He, and helium of mass 4, ^{4}He. In the 1960s, oceanographers observed that helium trapped in lava erupted from undersea volcanoes had a higher proportion of ^{3}He than observed in atmospheric helium. For decades the origin of deep-Earth helium was a great mystery: The ^{4}He was no surprise, because the alpha particles from natural radioactive decay become ^{4}He; the ^{3}He was the mystery because there was no known deep-Earth production mechanism that could explain the amount of ^{3}He observed. The explanation proffered for decades was that the ^{3}He is primordial, trapped since Earth's formation and mixed with just the right amount of ^{4}He to match observations. Based upon that explanation, a cottage-industry had developed making models of mantle degassing. But then along came the Oak Ridge National Laboratory fission product results for helium, which explained the helium as a byproduct of the georeactor, one of the intrinsic WEDD components.

When a uranium nucleus undergoes fission, it generally splits into two roughly equal, large fragments. This is called binary fission. But, once in every 10,000 fission events, the nucleus undergoes ternary fission and splits into three pieces, two large and one very small. Tritium, hydrogen of mass 3, ^{3}H, is a prominent very small fragment. Tritium is radioactive and decays to become ^{3}He. From the Oak Ridge calculations, the fission products ^{3}He and ^{4}He occur in the same ratio of values as observed in oceanic lava. This is strong evidence that the georeactor exists and is the source of the deep-Earth helium because the range of georeactor-produced helium isotope ratios match quite precisely geologically measured values, as we shall presently see.

The Oak Ridge National Laboratory nuclear calculations are state-of-the-art and quite sophisticated. So, for publication we decided on the *Geophysical Journal International* and submitted our manuscript to a French editor, imagining that, because France has one of the most advanced nuclear power industries, there would be a large pool of capable referees. Typically, upon submitting an article to a journal, the communicating author receives an acknowledgement of receipt and a manuscript identifying number. After receiving no response, we waited, sent email messages, even left a phone message, but still got no reply. Finally, we contacted the journal's London office and learned that the French editor had never logged in our submission. Eventually, presumably under pressure from the London office, the French editor obtained pejorative reviews from several referees, and rejected our manuscript based upon those reviews, which contained no scientifically valid criticisms. From the referees' remarks, clearly none had a background in the physics of nuclear reactors, which is strange because the manuscript was all about nuclear reactor physics!

Hatten S. Yoder, Jr., was the Director Emeritus of the Geophysical Laboratory of the Carnegie Institution of Washington, DC. Yoder was one of those rare individuals who fearlessly challenged geological orthodoxy and succeed – quickly, not in geological time as sometimes seems to be the case, for example, Alfred Wegener [12]! Among his many accomplishments, in 1993, Yoder published a "Timetable of Petrology," his list of fundamental geoscience discoveries. It begins with Strabo in 7 AD who "Recognized basalt from Etna Volcano had solidified from melt. Related earthquakes to volcanic eruptions." [13]. Among the contributions in his "Timetable of Petrology" Yoder lists the following:

- **1979 Herndon   Inner core of earth made of nickel silicide.**
- **1980 Herndon   Core and lower mantle of the earth are chemically analogous to the two components of enstatite chondrite meteorites on the basis of mass ratios of the elements.**

When Yoder heard about the French debacle, he was aghast, and offered to communicate our manuscript to the *Proceedings of the National Academy of Sciences, USA*. He obtained the required reviews; the paper was accepted, and it was published in 2001 [14].

The editor of *Discover* magazine noticed our *National Academy of Sciences* article and brought it to the attention of science writer Brad Lemley. After a thorough investigation, Lemley penned "Nuclear Planet," the feature article, the cover story of the August 2002 issue of *Discover* magazine. The article was masterfully written; precisely describing the step-by-step logical progression of understanding that began with my 1979 nickel silicide inner core publication and culminated with the georeactor. At long last, albeit in a popular magazine, my ideas had become available. Scientists and those interested in science throughout the world were exposed to WEDD, although I had not yet coined that term to refer to the logically connected explanations of Earth's origin, composition, decompression, development of georeactor and magnetic field, mountain building and so on.

In Switzerland, Walter Seifritz, a senior nuclear reactor engineer, verified my calculations [15]. In India, K. R. Rao, Associate Editor of *Current Science*, wrote a well-documented *Research News* article, entitled "Nuclear Reactor at the Core of the Earth! - A Solution to the Riddles of Relative Abundances of Helium Isotopes and Geomagnetic Field Variability" [16]. For a brief time I entertained the notion that the resultant public awareness would cause academicians to examine WEDD and compare it to the standard model.

There are many different types of manmade nuclear reactors, some, for example, employ fast-moving neutrons to produce (breed) reactor fuel, others use slow neutrons; design configurations vary greatly. For more than thirty years, scientists and engineers at Oak Ridge National Laboratory have developed and continuously refined software for computer simulations of the operation of the various types of nuclear reactors. The software has been validated by comparing the computer predictions for a particular nuclear reactor with actual reactor operating data and with analyses of its spent fuel, where its fission products are mostly trapped. The fission product data from our original georeactor calculations, as with those from a manmade reactor, were collected together as with other fission products over the time of the run, *i.e.*, they were continuously added together just as they would collect in nuclear reactor fuel. I saw the need to have instantaneous values, collected individually for each time increment and not added together. As I had not yet acquired the licensed software, with hat in hand I asked the good people at

Oak Ridge to make a few specific runs and collect instantaneous fission product data for helium isotopes. Graciously, they complied. Figure 5.6 presents the Oak Ridge National Laboratory instantaneous helium fission products, expressed as ^{3}He to ^{4}He ratios relative to that ratio in air. It is sometimes said that good data "talk to you." These data told a fascinating story which shed new light on the meanings of helium in geological measurements and their connection to WEDD.

Observe in Figure 5.6 that throughout most of Earth's lifetime, the georeactor ^{3}He to ^{4}He ratio increases slowly over time and has a range of values similar to those observed in basalt from along the mid-ocean ridges. But, look. Suddenly, more-or-less near the present time, the ^{3}He to ^{4}He ratio increases dramatically as the georeactor begins to run out of fuel. High helium ratios are observed in basalt from the Hawaiian Islands and from Iceland; these high values portend the demise of the georeactor. According to these calculations, perhaps one billion years from now, perhaps one hundred years, the georeactor will run out of fuel. Since, according to WEDD, the georeactor is the energy source (and production mechanism) for the geomagnetic field, if this happens during time of our technology-based society, the effects would include global disruption of telecommunications and planet-wide aurora borealis – a rather pretty apocalypse. No one knows, but sound scientific investigations should make it possible to project more precisely the lifetime of the deduced georeactor.

I described the new Oak Ridge National Laboratory helium data and explained its geophysical implications in the manuscript, entitled "Nuclear Georeactor Origin of Oceanic Basalt ^{3}He/^{4}He, Evidence, and Implications." Hatten Yoder obtained reviews and communicated the manuscript to the *Proceedings of the National Academy of Sciences, USA*. As this was a follow-on and an extension of the 2001 paper, we expected it to sail through the editorial process. Instead, after a time I learned that the manuscript would be subjected to further reviews by three anonymous referees who were members of National Academy of Sciences [NAS]. That is an unusual circumstance, but permitted by NAS policy.

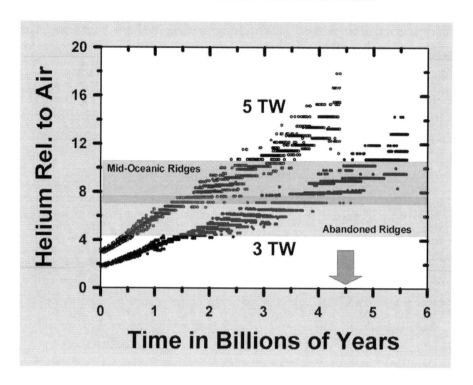

Figure 5.6 Fission product ratio $^3He/^4He$, relative to that of air, R_A, from nuclear georeactor numerical calculations at 5 terawatts, TW, (upper) and 3 TW (lower) power levels [17]. The band for measured values from mid-oceanic ridge basalts is indicated by the solid lines. The age of the Earth is marked by the arrow. Note the distribution of calculated values at 4.5 billion years, the approximate age of the Earth. The increasing values are the consequence of uranium fuel burn-up. Icelandic deep-Earth basalts present values ranging as high as 37 times the atmospheric value [37].

When I received the reviews, I was not delighted. Although it was clear the referees wanted the reject and eviscerate my work, they failed to make a single, scientifically valid criticism. One NAS referee, for example, wrote, "I hope the author is willing to ... proposed tests, rather than a definitive explanation of details of magma geochemistry." What's wrong with that comment aside from the grammar? The manuscript had nothing whatsoever to do with "magma geochemistry." Another NAS referee without explaining why wrote, "I don't believe the geophysical conclusion and the assertive tone are inappropriate." The third referee used only one word sentences. Bear in mind that these referees are members of one of the world's most prestigious scientific societies, reviewing an article for publication in that organization's premier scientific journal. So, foolishly as usual, I addressed the referees' comments. Instead of accepting the manuscript, the editor sent it back to the same referees who then responded with more comments, again lacking scientifically valid criticisms.

Yoder said he "smelled a rat." He began to make inquiries. What he found in essence was this: The editor, originally intending to publish the paper, sent a copy to a NAS member asking if he would like to write an accompanying commentary. Unfortunately, he chose the one NAS member who had a serious conflict of interest: This member's own speculative idea about the origin of deep-Earth helium was that ^{3}He arrived from space on cosmic dust grains which were then subducted into the mantle by assumed mantle convection. Instead of writing a commentary, that NAS member convinced the Editor-in-Chief, a biologist, that more reviews were needed and, of course, he would recommend NAS referees. When the Editor-in-Chief learned that editorial integrity had been compromised, he accepted the paper for publication [17].

What are the lessons to be learned from this fiasco? By an Act signed by U. S. President Abraham Lincoln (1809-1865), the National Academy of Sciences is charged with providing scientific advice to the U. S. Government. What quality of advice can the U. S. Government expect from an organization whose members act to suppress science? Membership in the National Academy of Sciences ought to reflect the highest scientific achievements together with the highest integrity; in some instances it does. But in too many instances membership is based upon popularity and political considerations. Some members are elected to the National Academy of Sciences who never made important fundamental scientific discoveries. And what about the NAS members who tried to suppress my manuscript? I complained about their behavior to the NAS President, who to my knowledge did nothing. Where were their ethics, their integrity? If gold rusts, what will iron do? Science is all about integrity. From long experience I have observed that those who deceive others end up deceiving themselves, cheating themselves and others from opportunities for making important scientific discoveries.

So, after all this, what were the consequences after publication of my georeactor-helium origin? Making mantle degassing models ceased, but to my knowledge no one supported by the National Science Foundation or NASA ever cited my work in that context.

Further Evidence of Whole-Earth Decompression Dynamics (WEDD)

The identity of georeactor-produced helium ratios with those ratios geologically observed in oceanic basalt is strong evidence for georeactor existence. But there is more, more georeactor evidence and further geophysical implications once we adopt the perspective shift, once we cut through the coprolitic strata of official geoscience which has, to use another geo-metaphor, ossified reason with its:

- **Circa 1940 ordinary chondritic Earth composition,**
- **Standard model, codified just two decades later,**
- **Earth of unchanging dimension,**
- **Continent parts colliding, sticking together, and breaking up again, and again, repeatedly, and**
- **Surface components in seeming perpetual motion driven by unseen, unproven mantle convection.**

However speculative the world of WEDD might seem at first glance, it meets better, more naturally, with fewer *ad hoc* additions, and presents a logical, "geochemically correct" (if I may coin that non-pejorative phrase) view of our Earth. The standard model, based on demonstratively false presuppositions, is more familiar but fictive. It stands alone, a logical disconnect from what many consider the grand unification of modern geology, plate tectonics, which itself stands alone and apart from the age-old question of the origin of mountains characterized by folding, such as the Alps, and from the primary origin of fjords and submarine canyons. Moreover, it is without clearly delineated energy sources: Radioactivity-produced heat, insufficient to match heat presently exiting Earth, is assumed to be augmented by ill-defined "heat left over from planetary formation." These disconnects are like the descriptions of an elephant given by blind men, each giving tactile examination to a different selected body-part, according to the ancient Indian parable. (Figure 5.7) Nature produces, not a set of disconnected body-parts, but the complete elephant; Whole-Earth Decompression Dynamics describes the whole elephant.

There are two main energy sources for powering geological activity that are consequences of Earth's early formation as a Jupiter-like gas giant:

- **Stored energy of protoplanetary compression, and;**

- **Georeactor nuclear fission and radioactive decay energy.**

Heat from the former seems conspicuously evident in the volcanic chain of mid-ocean ridges that encircle the globe like stitching on a baseball. Georeactor heat, identified in lava outpourings by its signature traces of helium, drives the formation of island chains, called "hotspots," and the massive eruptions of flood basalts, called "Traps."

A global, intimate connection exists between these two great energy sources. The Hawaiian Islands and the related Emperor seamount chain sequentially formed as oceanic crust moved across a persistent source of heat, a hotspot, as first suggested in 1963 by Wilson [18]. Some 49 hotspots have been identified, but not all have identical surface or compositional manifestations; some may have shallow heat sources. About 20%, though, including the Hawaiian Islands and Iceland, are characterized by basalts containing helium with high ^{3}He/^{4}He ratios (greater than ten times the corresponding ratios in air) that, I posit, are the signatures of georeactor-produced heat. Helium with these high ^{3}He/^{4}He ratios have been measured in massive flood-basalts, such as in the Siberian Traps which erupted 250 million years ago. They provide evidence, I believe, of their georeactor-powered hotspot origin even though the hotspots themselves are long gone.

Hotspots power the volcanic activity that is continuing to produce basalt-lava that forms the Hawaiian Islands and Iceland. Seismic tomography appears to image vertical, column-like "low velocity zones," sometimes called "mantle plumes," extending to the edge of the core for each of those two hotspots which, of course, is wholly consistent with the georeactor origin of that heat. That heat transport, however, cannot represent matter transport by convection because the mantle is too bottom-heavy and is not devoid of friction [19].

Figure 5.7 In the ancient Indian parable, a number of blind men attempt to describe an elephant as they touch particular parts of its body. The blind man who touched only its leg said, "It is like a pillar." Each proffered a different description based upon the body-part touched. This has been the way of the standard model, plate tectonics, and ordinary-chondritic Earth composition. Figuratively speaking, Whole-Earth Decompression Dynamics begins as the description of the "elephant" from which each particular "body-part" – energy sources, composition, geomagnetic field, separation of continents, seafloor topology, fold-mountain formation, etc. – can be deduced.

Water, uniformly distributed upon soil, often percolates downward by gravity in a non-uniform way, forming channels through paths of less resistance. An analogous process, I have suggested [19], might occur in the Earth's mantle for the upward-channeling of heat. Innumerable-layers of buoyancy-driven micro-convection in conjunction with thermal conduction in the density layered mantle perhaps operates to directionally-bias and/or augment the flow of core-derived heat upward. Helium, being a small, light, fast-moving noble gas, presumably is able to make its way upward toward regions of lower density along these heat-channels.

Hotspot formation beneath continental-rock surface and localized swelling produced by sub-surface magma augments continent-fragmentation which provides increased surface-area to accommodate decompression-driven increasing planetary volume. The stress of whole-Earth decompression along with the appearance of an underlying hotspot appears to be the combined agent involved in continent fragmentation and the concomitant opening of new ocean basins. A georeactor-powered hotspot underlies the current break-up of Africa along the East African Rift System and appears to have contributed to the opening of Gulf of Eden and the Red Sea. Similarly, the opening of the North Atlantic Ocean 61 million years ago appears to be associated with the Iceland hotspot.

But how universal is the association of continent-fragmentation with georeactor-produced heat? The details are not entirely clear; there is much yet unknown [20].

Variable Heat Output, Raghavan and the Antineutrino Test

Recently, the Norwegian scientists Mjelde and Faleide [21] discovered that basalt eruptions in the Hawaiian Islands and in Iceland varied significantly over time. Remarkably, the pulse-like variations in time and productivity in each case were synchronized, as if being orchestrated by a common mechanism, even though the two hotspots are located on opposite sides of the globe. Subsequent work by Mjelde, Wessel, and Müller [22] suggests the co-pulsations are a global hotspot phenomenon. The commonality appears to represent changes in heat from the Earth's core. Georeactor-heat produced by nuclear fission can be variable, unlike heat from the natural decay of long-lived radioactive isotopes, which is essentially constant, decreasing slightly over very-long periods of time.

The discovery [23], presented in Chapter 1, that number earthquakes of magnitude ≥ 6 have been increasing, at least since 1973, is further evidence of Whole-Earth Decompression Dynamics and its potential for geophysical variability. As noted Shearer and Stark [24], from the standpoint of the standard model/plate tectonics: "Moreover, no plausible physical mechanism predicts real changes in the underlying global rate of large events." But with WEDD there is a physical mechanism.

For decades, geophysicists have assumed that heat transport within the Earth takes place via conduction, convection, or by radiation, the latter of which is applicable only to the surface and atmosphere. I have introduced two additional heat transport mechanisms, namely, mantle decompression thermal tsunami [25] and heat channeling [19]. The decades-old idea that heat exiting the Earth is essentially constant, changing only over extremely long periods of time and then only decreasing, is not valid, yet nevertheless remains one of the basic (flawed) assumptions underlying climate models.

As early as 1930, it seemed that energy mysteriously disappeared during the process of radioactive beta decay. The energy account sheet simply did not balance. To preserve the idea that energy is neither created nor destroyed, "invisible" particles were postulated to be the agents responsible for carrying energy away unseen. Finally, in 1956 these "invisible" antineutrinos were detected experimentally. Later, neutrinos from the Sun were detected as well as from a supernova. Antineutrinos were detected from nuclear fission reactors. It is not surprising then that R. S. Raghavan, a neutrino expert, after learning about the georeactor during a lunch-time seminar at Bell Laboratories, would author a paper, entitled "Detecting a Nuclear Fission Reactor at the Center of the Earth" [26]. Raghavan showed that antineutrinos resulting from the products of nuclear fission have a different energy spectrum than those resulting from the natural radioactive decay of uranium and thorium.

As early as the 1960s, there was discussion of antineutrinos being produced during the decay of radioactive elements in the Earth. In 1998, Raghavan was instrumental in demonstrating the feasibility of their detection [27]. Now, Raghavan's paper on detecting a deep-Earth nuclear fission reactor stimulated intense interest worldwide, especially with groups in Italy, Japan and Russia. Russian scientists expressed well the importance: *"Herndon's idea about georeactor located at the center of the Earth, if validated, will open a new era in planetary physics"* [28].

Antineutrinos, sometimes called geoneutrinos, can literally fly through the Earth interacting very, very little, if at all, with the matter of Earth, which is the reason they are difficult to detect. Antineutrino detection requires large, extremely sensitive detectors operated for long periods of time and buried deep underground to shield from cosmic rays. To date, detectors at Kamioka, Japan and at Gran Sasso, Italy have detected antineutrinos coming from within the Earth. After years of data-taking, the first measurements have been made. From the total energy output of uranium and thorium, estimated from deep-Earth antineutrino measurements, an upper limit on the georeactor nuclear fission contribution was determined to be either 26% (Kamioka, Japan) [29] or 15% (Gran Sasso, Italy) [30]. The total georeactor contribution may be somewhat greater, though, as some georeactor energy comes from natural radioactive decay as well as from nuclear fission. Clearly, the antineutrino data were unable to refute georeactor existence and, in fact, provide evidence in support of the terracentric nuclear fission georeactor. For the future, improving antineutrino detection efficiency and lowering background will be beneficial, but will not delineate uranium and thorium locations; for that, a new generation of detectors will be required, detectors that not only signal an antineutrino detection event, but sense its direction.

There is a back-story, a dark side to the above discussion of antineutrino evidence that is important to present as it pertains to and sheds light on the institutional, compromised position of university physicists and geophysicists. A scientist in Europe told me that Raghavan had told him that his paper on detecting a nuclear fission reactor at the center of Earth had been rejected by two journals, *Physical Review Letters* and *Physics Letters*, because – I am paraphrasing here – one or more undisclosed reviewers objected to my georeactor concept. To the European, the implied warning seems clear: Cite my work and your own papers might not get published. A more generous view of human nature might conclude I'm being paranoid here, but the above is probably a more logical conclusion, especially when considered in light of the historical record of geoneutrino publications.

Antineutrinos can fly through the Earth virtually unimpeded. Although vast numbers of antineutrinos are produced, very, very few can be detected. Detection is the major challenge; huge, extremely sensitive detectors are required. Currently neutrino and antineutrino physics is a "hot" area for investigation, holding the promise of potentially important discoveries, and there are groups throughout the world whose detection systems are in various stages of development. In the area of antineutrino detection, the U. S. – Japan consortium, referred to by the acronym KamLAND, is technologically at the vanguard.

In July 2005, in a paper published in *Nature*, the KamLAND consortium reported the first detection of antineutrinos originating from within the Earth [31]. But what the paper said and what it should have said, I may be able to convince you, are two entirely different things. In easy to understand terms, in my view, this is what the paper should have said:

In just over two years of taking data, a total of 152 "detector events" were recorded. After subtracting for the background from commercial nuclear reactors and making corrections for contamination, only 20-25 "detector events" were considered to be from antineutrinos originating within the Earth. Within the limitations of the experiment, it is absolutely impossible to ascertain the proportion of those that may have resulted from the radioactive decay of uranium and thorium, or may have been produced from a nuclear fission georeactor at the center of the Earth.

Instead, what the 87 authors of the KamLAND consortium did was to mislead the scientific community and the general public by wholly and intentionally ignoring the possibility of georeactor-produced antineutrinos. Raghavan's 1998 paper on measuring the global radioactivity in the Earth was cited [27], but not his 2002 paper "Detecting a Nuclear Fission Reactor at the Center of the Earth" [26]. And, there was absolutely no reference to any georeactor paper.

The KamLAND misrepresentation was undergirded by a "News and Views" companion article in the same issue [32] that discussed radioactive decay heat production in the Earth, noting:

The remaining heat must come from other potential contributors, such as core segregation, inner-core crystallization, accretion energy or extinct radionuclides – for example the gravitational energy gained by metal accumulating at the centre of the Earth, which is converted to thermal energy, and the energy added by impacts during the Earth's initial growth.

Absolutely no mention was made of georeactor-produced heat, which is on a firmer scientific foundation than some of the "other potential contributors" mentioned.

So, how stands the U. S. Department of Energy (DOE), which supported the KamLAND misrepresentation? Writing in response to my complaint, the

Director of the DOE Office of Science, on behalf the Secretary of DOE, stated this:

> **The following disclaimer is implicit for all scientific publications of research sponsored by the Department: … Neither the United States Government or any agency thereof, nor any of their employees, makes any warranty, express or implied, or assumes any legal liability or responsibility for the accuracy, completeness, or usefulness of any information, apparatus, product, or process disclosed ….**

This is a clear admission that the U. S. Government is not legally obligated to tell the truth, but does that come as a surprise?

For Japan, the first detection of antineutrinos from within the Earth by the KamLAND consortium, a very difficult technological triumph, should rightly have been cause for celebration. But instead of telling the full truth, namely, acknowledging the possibility that some geo-neutrinos might be georeactor-produced, Japanese KamLAND scientists dropped the ball and became party to the same American anti-science behavior and in doing so dishonored themselves and Japan.

Curiously, all that was really required in their paper was one carefully worded sentence with appropriate references. I complained to Japan's Minister for Science and Technology and to their credit, Japanese KamLAND scientists thereafter rightly cited potential georeactor contributions. Some American physicists and geophysicists, however, continue to fail to cite georeactor publications, and they do so with the tacit approval of their university officials (to whom I had complained), the ones who sign for millions of dollars of U. S. Government contracts and grants to conduct scientific research, for examples, see [33, 34].

When I encounter a new idea in the scientific literature, I ask myself, "suppose that idea is correct. If so, what does it mean? What are the potential consequences? What discoveries might lie ahead?" To me, truth is paramount. Truth is what science is all about. It may be relative, it may be uncertain, but the scientists' obligation is to tell the truth, the full truth; that is the point of being a scientist – in itself a noble endeavor. From long experience I have observed that those who do not tell the full truth in science impede subsequent discoveries, their own and those of others.

Peer Review

Nobel Laureate Sir Derek Barton once told me that upon occasion an unscrupulous editor would pretend to be a reviewer so as to generate a negative review so that he/she would have a basis for suppressing the paper.

The thought underlying the policy of editors selecting referees is that editors cannot be expected to possess the broad scientific background to make reasoned judgment; referees may provide the necessary broader-based expertise. That policy has changed. Editors at major journals, including those published by Copernicus, Elsevier, the Nature Company, and Springer, allow their editors to reject scientific submissions without review, without referees' comments. What's wrong with that policy change? In a broadest sense it does not take into account the nature of human behavior: When individuals are confronted with new ideas and observations their responses often leave much to be desired. In 1623, Galileo, one of the greatest scientists of the millennium, precisely characterized human response to new ideas in a letter written to Don Virginio Cesarini (translated by Stillman Drake, page 231) [35]:

> I have never understood, Your Excellency, why it is that every one of the studies I have published in order to please or to serve other people has aroused in some men a certain perverse urge to detract, steal, or depreciate that modicum of merit which I thought I had earned, if not for my work, at least for its intention. In my Starry Messenger there were revealed many new and marvelous discoveries in the heavens that should have gratified all lovers of true science; yet scarcely had it been printed when men sprang up everywhere who envied the praises belonging to the discoveries there revealed. Some, merely to contradict what I had said, did not scruple to cast doubt upon things they had seen with their own eyes again and again....How many men attacked my Letters on Sunspots, and under what disguises! The material contained therein ought to

have opened the mind's eye much room for admirable speculation; instead it met with scorn and derision. Many people disbelieved it or failed to appreciate it. Others, not wanting to agree with my ideas, advanced ridiculous and impossible opinions against me; and some, overwhelmed and convinced by my arguments, attempted to rob me of that glory which was mine, pretending not to have seen my writings and trying to represent themselves as the original discoverers of these impressive marvels....I have said nothing of certain unpublished private discussions, demonstrations, and propositions of mine which have been impugned or called worthless....Long experience has taught me this about the status of mankind with regard to matters requiring thought: the less people know and understand about them, the more positively they attempt to argue concerning them, while on the other hand to know and understand a multitude of things renders men cautious in passing judgment upon anything new.

An editor's knee-jerk response to new ideas and observations might be ameliorated if required to consider reviewers' comments.

Editor rejection without reviews also opens the door to instantaneous science suppression by editors who might be associated with a malevolent agenda, such as revealed by the Climategate emails [36].

Consider the following example: In the twenty years that elapsed since publication of the first paper demonstrating the feasibility of a nuclear fission reactor at the center of the Earth as the energy source for the geomagnetic field, much development and understanding took place. As there had been no published review article on the subject, I wrote one and submitted it to the Elsevier journal *GeoResJ*. The assigned editor, a geology professor at the University of Oxford, with no training in the subject of nuclear reactor physics, rejected the review article, without referee reviews.

Here is a contrast in intellectual integrity: Following the *GeoResJ* rejection, I submitted the manuscript to *Current Science*, which since 1932 has been published in association with the Indian Academy of Sciences. In that instance, the editor sent the manuscript to knowledgeable referees who asked for clarification and who asked me to provide additional information, which I did. And it was published [2], entitled "Terracentric Nuclear Fission Georeactor: Background, Basis, Feasibility, Structure, Evidence and Geophysical Implications." (Appendix B)

The costs of scientific publication in most instances and in most journals are borne by taxpayers. Research grants/contracts typically provide funds to pay for the publication of research results and/or permit the publishers to charge, typically $40.00 for each electronic copy of an article. In so far as many journal editors have conflicts of interest and yet can reject scientific articles without referee reviews, in some cases, lamentably, taxpayers are paying for a system that lacks integrity, a system that may promote a politically-driven viewpoint to the exclusion of new ideas and the most recent research results; a potentially biased publication system. Having been rejected, without benefit of reviews, by journals published by Copernicus, Elsevier, the Nature Company, and Springer, I doggedly appealed to the Chief Executive Officers of Elsevier and Springer and to the Director of Publications of Copernicus. My complaints to the big name science publishers got nowhere. With specific reference to Copernicus and Springer, both German organizations, I even complained to the German Academy of Sciences Leopoldina (founded in 1652) of the risk of science suppression by editors no longer required to get independent reviews. But nothing came of it.

It seems to me that electronic publication, without gatekeepers and secret reviews, by a neutral, objective agency, such as by the Library of Congress, would be considerably more cost effective and less subject to science suppression. The corporate takeover of science publishing, while in keeping with other changes in globalization, presents us a new and serious obstacle in the pursuit of science which is arguably our species' most significant contribution.

The fictive idea of a parallel universe metaphorically applies to georeactor-related human activities. In the dark universe where "science" is an oxymoron and the methods of malevolent politics hold, georeactor progress is stymied by misrepresentation and suppression. In a parallel universe of light, the long-held standards of science prevail, and scientific advances take place. Recently, for example, Kao-Ping Lin, Rong-Jiun Sheu, and Shiang-Huei Jiang of the National Tsing Hua University, Taiwan, Republic of China, reported research advances using the new Oak Ridge software: Their lengthy scientific article, "Actinide Inventory in Herndon's Georeactor Operating throughout Geologic Time," has been accepted for publication in *Annals of Nuclear Energy* [38].

Chapter 5 References

1. Domogatsky, G.V., Kopeikin, V. I., Mikaelyan, L. A., and Sinev, V. V., *Neutrino geophysics at Baksan I: Possible detection of georeactor antineutrinos* Physics of Atomic Nuclei, 2005. 68(1): p. 62-72.

2. Herndon, J.M., *Terracentric nuclear fission georeactor: background, basis, feasibility, structure, evidence and geophysical implications.* Current Science, 2014. 106(4): p. 528-541.

3. Bacquerel, H., *Sur les radiations emises par phosphorescence.* Comptes Rendus, 1896. 122: p. 501-503.

4. Hahn, O. and F. Strassmann, *Uber den Nachweis und das Verhalten der bei der Bestrahlung des Urans mittels Neutronen entstehenden Erdalkalimetalle.* Die Naturwissenschaften, 1939. 27: p. 11-15.

5. Fermi, E., *Elementary theory of the chain-reacting pile.* Science, Wash., 1947. 105: p. 27-32.

6. Kuroda, P.K., *On the nuclear physical stability of the uranium minerals.* J. Chem. Phys., 1956. 25(4): p. 781-782.

7. Kuroda, P.K., *On the infinite multiplication constant and the age of uranium minerals.* J. Chem. Phys., 1956. 25(6): p. 1295-1296.

8. Murrell, M.T. and D.S. Burnett, *Actinide microdistributions in the enstatite meteorites.* Geochim. Cosmochim. Acta, 1982. 46: p. 2453-2460.

9. Herndon, J.M., *Feasibility of a nuclear fission reactor at the center of the Earth as the energy source for the geomagnetic field.* Journal of Geomagnetism and Geoelectricity, 1993. 45: p. 423-437.

10. Herndon, J.M., *Planetary and protostellar nuclear fission: Implications for planetary change, stellar ignition and dark matter.* Proceedings of the Royal Society of London, 1994. A455: p. 453-461.

11. Herndon, J.M., *Sub-structure of the inner core of the earth.* Proceedings of the National Academy of Sciences USA, 1996. 93: p. 646-648.

12. Yoder, H.S., Jr. and C.E. Tilley, *Origin of basalt magmas: an experimental study of natural and synthetic rock systems.* Journal of Petrology, 1962. 3: p. 342-532.

13. Yoder, H.S., Jr., *Timetable of Petrology.* J. Geol. Ed., 1993. 41: p. 447-489.

14. Hollenbach, D.F. and J.M. Herndon, *Deep-earth reactor: nuclear fission, helium, and the geomagnetic field.* Proceedings of the National Academy of Sciences USA, 2001. 98(20): p. 11085-11090.

15. Seifritz, W., *Some comments on Herndon's nuclear georeactor.* Kerntechnik, 2003. 68(4): p. 193-196.

16. Rao, K.R., *Nuclear reactor at the core of the Earth! - A solution to the riddles of relative abundances of helium isotopes and geomagnetic field variability.* Current Science, 2002. 82(2): p. 126-127.

17. Herndon, J.M., *Nuclear georeactor origin of oceanic basalt $3He/4He$, evidence, and implications.* Proceedings of the National Academy of Sciences USA, 2003. 100(6): p. 3047-3050.

18. Wilson, J.T., *A possible origin of the Hawaiian Islands.* Can. J. Phys., 1963. 41: p. 863-870.

19. Herndon, J.M., *Geodynamic Basis of Heat Transport in the Earth.* Current Science, 2011. 101(11): p. 1440-1450.

20. Courtillot, V., *Evolutionary Catastrophes*2002, Cambridge: Cambridge University Press.

21. Mjelde, R. and J.I. Faleide, *Variation of Icelandic and Hawaiian magmatism: evidence for co-pulsation of mantle plumes?* Mar. Geophys. Res., 2009. 30: p. 61-72.

22. Mjelde, R., P. Wessel, and D. Müller, *Global pulsations of intraplate magmatism through the Cenozoic.* Lithosphere, 2010. 2(5): p. 361-376.

23. Herndon, J.M., *Variable Earth-heat production as the basis of global non-anthropogenic climate change.* Submitted to Current Science, September 4, 2014.

24. Shearer, P.M. and P.B. Stark, *Global risk of big earthquakes has not recently increased.* Proc. Nat. Acad. Sci. USA, 2012. 109(3): p. 717-721.

25. Herndon, J.M., *Energy for geodynamics: Mantle decompression thermal tsunami.* Current Science, 2006. 90(12): p. 1605-1606.

26. Raghavan, R.S., *Detecting a nuclear fission reactor at the center of the earth.* arXiv:hep-ex/0208038, 2002.

27. Raghavan, R.S. and e. al., *Measuring the global radioactivity in the Earth by multidectector antineutrino spectroscopy.* Phys. Rev. Lett., 1998. 80(3): p. 635-638.

28. Domogatski, G.V., et al., *Neutrino geophysics at Baksan I: Possible detection of georeactor antineutrinos.* Physics of Atomic Nuclei, 2005. 68(1): p. 69-72.

29. Gando, A., et al., *Partial radiogenic heat model for Earth revealed by geoneutrino measurements.* Nature Geosci., 2011. 4: p. 647-651.

30. Bellini, G. and e. al., *Observation of geo-neutrinos.* Phys. Lett., 2010. B687: p. 299-304.

31. Araki, T., et al., *Experimental investigation of geologically produced antineutrinos with KamLAND.* Nature, 2005. 436: p. 499-503.

32. McDonough, W.F., *Earth sciences: Ghosts from within.* Nature, 2005. 436: p. 467-468.

33. Dye, S.T., Huang, Y., Lekic, V., McDonough, W.F., Sramek, O., *Geo-neutrinos and Earth models.* Physics Procedia, 2014. arXiv:1405.0192.

34. McDonough, W.F., Learned, J. G., Dye, S.T., *The many uses of electron antineutrinos.* Physics Today, 2012. 65: p. 46-51.

35. Drake, S., ed. *Discoveries and Opinions of Galileo.* 1956, Doubleday: New York. 301.

36. Costella, J., ed. *The Climategate Emails.* 2010, The Lavoisier Group: Australia.

37. Hilton, D.R., et al., *Extreme He-3/He-4 ratios in northwest Iceland: constraining the common component in mantle plumes*. Earth Planet. Sci. Lett., 1999. 173(1-2): p. 53-60.

38. Lin, K-P, Sheu, R.-J., Jiang, S.-H. *Actinide inventory in Herndon's georeactor operating throughout geologic time*. Annans of Nuclear Energy. In press.

6. On the Trail of Earth's Magnetic Field

Since the discovery of secret things and in the investigation of hidden causes, stronger reasons are obtained from sure experiments and demonstrated arguments than from probable conjectures and the opinions of philosophical speculators of the common sort; therefore to the end that the noble substance of that great loadstone, our common mother (the earth), still quite unknown, and also the forces extraordinary and exalted of this globe may the better be understood, we have decided first to begin with the common stony and ferruginous matter, and magnetic bodies, and the parts of the earth that we may handle and may perceive with the senses; then to proceed with plain magnetic experiments, and to penetrate to the inner parts of the earth. — *William Gilbert, 1600, De Magnete.*

No other manifestation of Earth has been as seemingly inexplicable as the origin of Earth's magnetic field. Albert Einstein is said to have considered its origin one of the most important unsolved problems in physics.

In 1993, when I published the first calculations demonstrating the feasibility of a natural nuclear fission reactor at the center of the Earth [1], I was thinking of it only as an energy source powering the mechanism that produces the geomagnetic field. But, with continuing development and understanding, by 2007 I realized that the georeactor may well be not just the energy source, but also the mechanism itself that generates Earth's magnetic field [2, 3]. (Appendix B) This new concept for Earth's magnetic field, as we shall see, carried over to magnetic field origination in other planets and large moons and overcomes some of previous difficulties related to small body size.

In 1939, Walter Elsasser began a series of scientific publications in which he proposed that the geomagnetic field is produced within the Earth's fluid core by an electric generator mechanism, also called a dynamo mechanism [4-6]. Elsasser proposed that convection currents in the Earth's electrically-conducting iron-alloy core, twisted by Earth's rotation, act like a self-sustaining dynamo, a magnetic amplifier, producing the geomagnetic field.

For decades, Elsasser's dynamo-in-the-core was generally considered to be the only viable means for producing the geomagnetic field. It was usually, and still is cited without question. Given my new perspective, and lack of other geoscientists participating in it, I took it upon myself to investigate the problem. To my surprise, I discovered a problem, not with his idea of a convection-driven dynamo, but with its location, its operant fluid, its requisite "seed" magnetic field, and its energy source [2, 3, 7-9].

Elsasser's geomagnetic field generator involves motion; a moving charge produces a magnetic field. It was natural that he would consider a region within the Earth that was known to be liquid, the core, the body of liquid iron alloy that comprised about one-third of the Earth's mass. Elsasser invoked the process of convection to keep the liquid in motion and originally suggested that the decay energy of a central mass of uranium would provide the necessary power. Geoscientists at the time, however, believed that Earth resembles an ordinary chondrite, and could find no reason uranium might exist in the core [10]; instead, they ultimately came to believe in the existence of a different energy source. Without evidence, they assumed that the inner core is growing, crystallizing from the fluid iron alloy core, and releasing heat in the process. For seventy years, since Elsasser's dynamo-in-the-core idea, members of the scientific community have swallowed the notion that convection "must" exist in the core. Millions of dollars have been spent modeling convection and its ramifications in the Earth's liquid core without anyone asking, the preliminary question: was it even possible?

Computational models are computer programs that begin with a known end result (that which is being modeled) and a set of arbitrary assumptions, which must lead to the end result. Sometimes new assumptions are added, sometimes the data are result-selected, meaning the researcher cherry picks the model-input he or she needs for the model to give the intended result. Models are sometimes useful, but in my view they are not science. Models typically yield whatever the model-maker wants, whether correct or not. Recall the great waste incurred by making mantle convection models without realizing that the underlying assumptions, such as absence of friction, are inconsistent with Earth's behavior. Models incorporating sustained thermal convection in the Earth's fluid core, like those of mantle convection, rely upon assumptions that are inconsistent with Earth's behavior.

As in the case of mantle convection, but to lesser degree, compression of the fluid core by the weight above causes the fluid core-bottom to be some 16% denser (heavier) than the top. The very small decrease in density, less than 1%, caused by thermal expansion is not enough to make a "parcel" of bottom core material light enough to float to the top of the core – the process needed begin circulation and thus convection, a most efficient means of heat transfer, and the source of motion needed for Elsasser's dynamo. There is, however, a more serious impediment to core convection that does not have a parallel with mantle convection. The bottom of the fluid core cannot remain hotter than the top, as required for sustained convection, because the fluid core is wrapped in the thick insulating rock-blanket (the mantle); heat cannot readily escape. Further, the "Rayleigh Number" [11], a calculation derived in 1916 to predict the onset of convection in a liquid thin film of *uniform* density, is inappropriate to justify the onset of convection for the compressed core of *non-uniform* density. Without convection, there can be no convection-driven dynamo in the Earth's fluid core [9].

There is yet another, not often recognized, problem with Elsasser's dynamo-in-the-core idea: The dynamo mechanism is a magnetic amplifier; it requires a "seed" magnetic field to amplify. I am unaware of any means of producing a seed magnetic field within the Earth's fluid core.

By contrast, the nuclear fission georeactor has all the components necessary for powering and generating the Earth's magnetic field [2, 3, 7-9]. And indeed, the mechanism works exactly by Elsasser's convection-driven dynamo mechanism except that the location is moved to a place where it will work. As illustrated in Figure 6.1, Earth's deduced georeactor has a sufficient energy source located at its gravitational bottom, its nuclear fission sub-core, and a region above the nuclear sub-core where sustained thermal convection is possible, its nuclear-waste sub-shell. I propose that heat produced by sustained nuclear fission at the bottom of the nuclear-waste sub-shell is carried to the top by thermal convection where it encounters and exchanges heat with a massive heat sink, the inner core, which is surrounded by an even more massive heat sink, the core. This arrangement assures that the bottom of the convective region will always be hotter than the top. That is quite unlike the widely presumed but improbable convection in the fluid core – which is wrapped in an insulating blanket. Moreover, radioactive beta-decay within the nuclear-waste sub-shell provides a plethora of electrons whose motion may form the requisite seed magnetic field for amplification by convection-driven dynamo action.

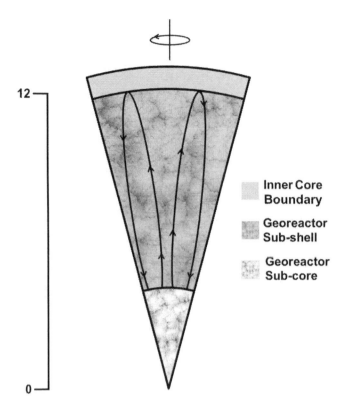

Figure 6.1 Schematic representation of the georeactor. Planetary rotation and fluid motions are indicated separately; their resultant motion is not shown. Stable convection is expected with adverse temperature gradient, *i.e.*, bottom hotter than top, and heat removal via heat sink at the top, *i.e.*, the inner core. Scale in kilometers.

When a rock cools from a high temperature, its magnetic minerals record the direction, and to some extent, the intensity of the prevailing magnetic field, analogous to a magnetic tape recorder or computer hard drive. Geologists take carefully oriented rock samples, then determine their direction of magnetization with a magnetometer and also determine the date of their last cooling. What they have found is that the geomagnetic field reverses from time to time. Over geologic time, magnetic North can switch and has switched to magnetic South and vice versa. The process happens irregularly. The average time between reversals of about 250,000 years is misleading. There are periods when reversals are more frequent, and periods of stability lasting as long as 50 million of years. The last reversal occurred about 750,000 years ago. Figure 6.2 shows the periods of normal and reversed magnetic polarity over the last 160 million years.

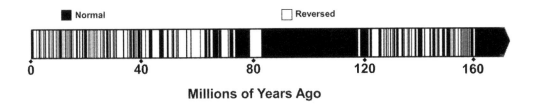

Figure 6.2 Geomagnetic polarity during the last 160 million years. Dark areas denote periods where the polarity matches today's polarity, called normal polarity; light areas denote periods where that polarity is reversed. Adapted from image courtesy of Atomie, based upon data from [18, 19].

Any concept that purports to describe the generation of the geomagnetic field should be able to explain the manner by which long-term stability is possible. It should also explain the physical basis for magnetic reversals. Long-term stable thermal convection does not seem possible in the fluid core, for reasons outlined above. Consequently, if the geomagnetic field is produced by Elsasser's convection-driven dynamo mechanism, the operant fluid must reside elsewhere. I suggested instead that the georeactor's nuclear-waste sub-shell is that location wherein the geomagnetic field is generated by long-term stable thermal convection.

Elsasser's dynamo is a magnetic amplifier, meaning that it requires a small magnetic "seed" field to strengthen and build into the large magnetic field. The seed magnetic field is particularly important during magnetic reversals. In the georeactor's nuclear waste sub-shell, where thermal convection is thought to take place, neutron-rich fission products decay predominantly by beta-decay – they shoot out electrons. Moving electrons produce magnetic fields; presumably these are the source of magnetic seed fields.

The natural configuration of the georeactor not only assures stable thermal convection, but stable nuclear reactor operation as well. The latter is particularly important as nuclear reactor fuel changes over time and, if unregulated, a nuclear reactor can run wild and consume all of its fuel. To understand georeactor self-regulation, consider this thought-exercise: Imagine that the georeactor nuclear sub-core starts to run wild. In its microgravity environment, the nuclear sub-core disassembles and mixes with the nuclear-waste sub-shell, dramatically diminishing the chain reaction. As the sub-core cools, uranium will settle out from the sub-shell and return to the sub-core which will cause the nuclear reaction to increase. Eventually, a state of dynamic equilibrium will be established between nuclear reactor output and uranium settling out, a natural self-regulation mechanism.

The idea of geomagnetic reversals arising as a consequence of dynamo-convection instability goes back to decades when dynamo-in-the-core convection erroneously was thought possible. No specific mechanism for causing that core-convection instability could be ascertained logically, though, as the mass of the fluid core is nearly one-third of Earth's mass. That is not the case for the georeactor which is considerable less massive. Reversals of the geomagnetic field are produced when stable convection is interrupted in the region where convection-driven dynamo action occurs, in the nuclear-waste sub-shell. Upon re-establishing stable convection, convection-driven dynamo action resumes with the geomagnetic field either in the same or a reversed direction.

The georeactor mass is less than one ten-millionth the mass of the fluid core. Consequently, two potentially de-stabilizing processes are possible, namely massive planet trauma and solar wind induction. Moreover, because of the georeactor's considerably smaller mass, reversals can occur much more quickly than previously thought.

The impact of a large asteroid or the violent splitting apart of continental land masses could conceivably de-stabilize georeactor dynamo-convection and cause a magnetic reversal. It is also possible that a particularly violent event on the Sun might lead to a reversal. The latter requires some explanation.

Geomagnetic Reversals

In the 19th century, Michael Faraday discovered that an electrical current could be transferred through a magnetic field, a process known as induction. Consider the apparatus in the schematic representation in Figure 6.3. In the diagram, the battery provides a source of direct current; an electric current is the flow of electric charge. Close the switch on the left and electric current flows in the left circuit. But only a spike, a momentary deflection of the meter in one direction is observed in the right hand circuit. Now, open the switch and a momentary deflection of the meter is again observed, but in the opposite direction. Induction is possible only with a *changing* electric current or a *changing* magnetic field.

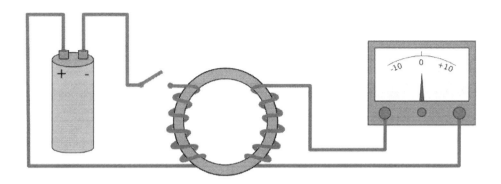

Figure 6.3 Schematic representation of Michael Faraday's discovery of induction, the transfer of electricity through a magnetic field. Close the switch on the left and electric current flows in the left circuit. But only a spike, a momentary deflection of the meter in one direction is observed in the right hand circuit. Now, open the switch and a momentary deflection of the meter is again observed, but in the opposite direction. Induction is possible only with a *changing* electric current or a *changing* magnetic field. Image courtesy of Eviatar Bach.

Earth is constantly bombarded by the solar wind, a fully ionized and electrically conducting plasma, heated to about 1 million degrees Celsius, that streams outward from the Sun and assaults the Earth at a speed of about 1.6 million kilometers per hour. The geomagnetic field deflects the brunt of the solar wind safely past the Earth, but some charged particles are trapped in donut-shaped belts around the Earth, called the Van Allen Belts. (Figure 6.4) Now the charged particles within the Van Allen Belts form a powerful ring current that produces a magnetic field that opposes the geomagnetic field. If the solar wind is constant, then the ring current is constant and no electric currents are transferred through the magnetic field into the georeactor by Faraday's induction. High-intensity changing outbursts of solar wind, on the other hand, will induce electric currents into the georeactor, causing ohmic heating, which in extreme cases, might disrupt dynamo-convection and lead to a magnetic reversal.

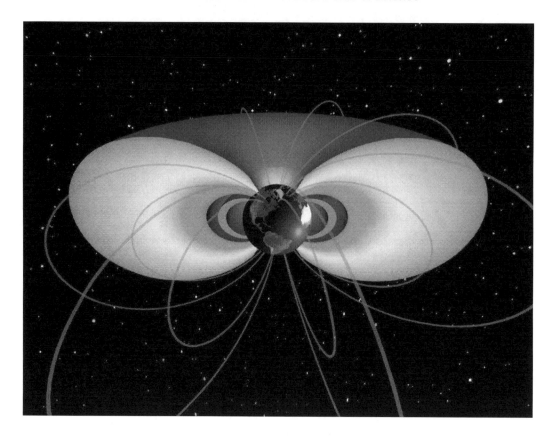

Figure 6.4 NASA artist's representation of Earth's Van Allen Belts wherein charged particles circulate as a ring current that produces a magnetic field in opposition to the geomagnetic field.

Magnetic reversals during the past 5 million years have occurred frequently, sometimes with short intervals between, as shown in Figure 6.5. Over the past 2,000 years the geomagnetic field has decreased in intensity about 35% to its present value, with nearly one-third of that decrease occurring in the last 150 years [12]. The Magnetic North Pole is currently drifting from Canada towards Siberia at a rate about four times as great as it did one hundred years ago. (Figure 6.5) So, what does this mean? Is a reversal imminent? Is the geomagnetic field approaching the end of its life? What is the timeframe? At present, the answers are not known. The best one can do at the present time is to set forth with truthfulness the relevant observations, and to seek objective understanding.

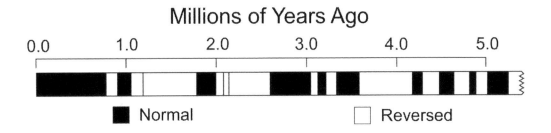

Figure 6.5 Geomagnetic polarity during the last 5 million years. Dark areas denote periods where the polarity matches today's polarity, called normal polarity; light areas denote periods where that polarity is reversed. Adapted from image courtesy of Intgr, based upon USGS figure.

In my opinion, there is much confusion in the scientific literature and in the popular media that stems from more than half a century of mistaken belief that the geomagnetic field originates in the Earth's fluid core: Traditional geoscience imagines a magnetic reversal process lasting over hundreds of years. Traditional geoscience also expects our magnetic field to hold up for another 4½ billion years, a figure which is based on the slow decay of natural long-lived radioactive elements. In my view, this thesis is old news, and erroneous news at that.

From ancient lava flows, scientists recently confirmed evidence of episodes of rapid geomagnetic field change – six degrees per day during one reversal and, for another reversal, one degree per week – were reported [13, 14]. A just-published study of a different reversal yielded geomagnetic field direction change of more than two degrees per year, meaning the full 180 degree reversal took place over a period of time comparable to a human lifetime [15]. Indeed, the relatively small mass of the georeactor is consistent with the possibility of a magnetic reversal occurring on a time scale as short as one month or several years.

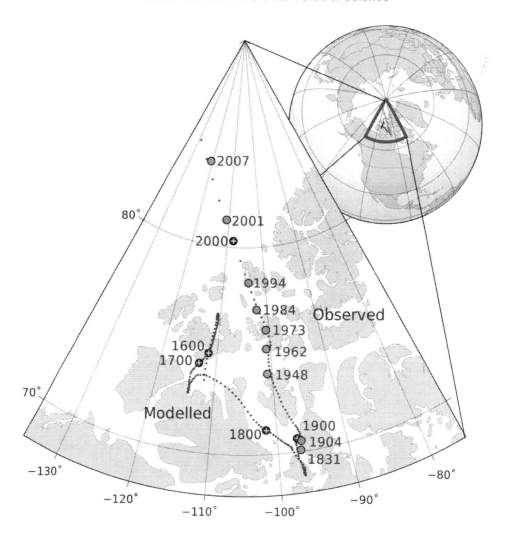

Figure 6.6 Magnetic north pole positions of the Earth. Poles shown are dip poles, defined as positions where the direction of the magnetic field is vertical. Solid circles mark magnetic north pole positions as determined by direct observation; circles with + are modelled positions. Image courtesy of Tento, based upon data from [20, 21].

Georeactor Extinction

A point of difference between WEDD and traditional geoscience involves my specialty, nuclear chemistry. Nuclear fission consumes uranium at a much faster rate than consumption by natural radioactive decay. The Oak Ridge National Laboratory georeactor computer simulations show that

^{3}He/^{4}He increases markedly as the uranium fuel tank approaches empty [16]. Interestingly, correspondingly high helium ratios are observed in lava erupted in Iceland and in Hawaii. It therefore appears that the georeactor is approaching the end of its lifetime and that, when it dies, the geomagnetic field will collapse soon thereafter. But when will that be? In a few years? In a few centuries? Longer? When the geomagnetic field collapses, will it restart as a reversal or will it never restart? The uncertainty is great.

There is a commonality in our Solar System's planet-formation processes, including planet-centered nuclear fission reactors [7, 8, 17]. Currently active internally generated magnetic fields have been detected for six planets (Mercury, Earth, Jupiter, Saturn, Uranus, and Neptune) and in at least one satellite (Jupiter's moon Ganymede). Magnetized surface areas of Mars and the Moon indicate the former existence of internally generated magnetic fields in those bodies, but the Martian and Lunar "georeactors" have long since ceased to generate magnetic fields. Venus is an enigma: Is the absence of a Venusian magnetic field the result of its low rotation rate or has its "georeactor" died? Earth's georeactor will eventually die and the geomagnetic field will collapse. When it does, it will be a catastrophe of Biblical proportions to our technology-based civilization and perhaps (but I hope not) also, due to agenda-driven science, humanity will still be unprepared.

At the time of the last reversal of the geomagnetic field, *homo erectus* had just begun to stand upright in Africa and may not yet have learned to put fire to use. Much has changed. Except in extremely remote regions, such as deep in the Amazonian rainforest, civilization has become electrified, electronically interfaced and interlaced; satellites are everyday intermediaries. Solar flares, coronal mass ejections from the Sun, occasionally overwhelm the geomagnetic shield and induce their own electric current into power grids. These intrusions are minuscule in comparison with what will happen during a complete collapse of the geomagnetic field, but they provide a glimpse of what is to be. For example, a large geomagnetic storm occurred March 13-14, 1989. This led to the collapse of the Canadian Hydro Québec power grid for a period of hours and two of its transformers sustained damage. The storm also damaged a transformer at a nuclear power plant in New Jersey and is suspected to have damaged two transformers in England.

Even the most severe solar-flare-caused geomagnetic storm is a momentary, recoverable event. When the geomagnetic field collapses, on the other hand, large segments of the population will be without electricity and without the prospect of rapid recovery. Electrical power grids, once used solely to convey and distribute electricity to cities and neighborhoods, will become themselves uncontrolled generators as the changing, electrically charged flux of the rampaging solar wind sweeps past, inducing into power lines suicidal bursts of electric current that will damage, short-circuit and destroy essential elements of power grids. Powerful, equipment-wrecking electrical currents will likewise be induced in gas and oil pipelines. Electric charges will build up on surfaces everywhere and reach staggeringly high potentials at edges and sharp points, posing risks of electrocution and igniting fires. The intense, continuous bombardment by the solar wind will fry the electronics of satellites, make radio communications impossible, disrupt navigation systems, and interfere with aircraft electronics. In short, modern infrastructure will become disconnected; our technology-based society will be instantly set back two hundred years, without the means to cope. Some may enjoy a brief aesthetic reprieve as aurora borealis patterns become heightened by the solar wind, which is no longer deflected around and past Earth as it once was when our magnetic shield was robust. But in general humanity will be in bad shape. The ancient evils will begin to re-emerge; disease, starvation, and pestilence. The struggle to survive will collapse social order; the barbarians will return. When will all this happen? Suddenly, next summer? Next century? Next millennium? What should we do?

Rather than making models based upon arbitrary assumptions, scientists should focus on making discoveries that lead to a better understanding of deep-Earth behavior, especially, to obtain more precise indications of the current state of activity within the georeactor. Knowing *when* is all-important to mobilizing preparedness; presently, there is too much uncertainty. Futurists and planners and macro-engineers might delight in the challenge of a lifetime, designing for a time when the Earth's magnetic field collapses and geomagnetic storms are daily occurrences. To prepare adequately for that, however, scientists need to extricate themselves from their present, unduly solidified mindset. On a more positive note though, I believe that, with sufficient warning, when Earth's magnetic field collapses, humanity will come together as never before to confront this common mega-peril. Perhaps it will be the wakeup call that makes our self-centered, pillaging species into a more mature and long-lasting, if less populous planetary civilization.

Chapter 6 References

1. Herndon, J.M., *Feasibility of a nuclear fission reactor at the center of the Earth as the energy source for the geomagnetic field.* Journal of Geomagnetism and Geoelectricity, 1993. 45: p. 423-437.

2. Herndon, J.M., *Nuclear georeactor generation of the earth's geomagnetic field.* Current Science, 2007. 93(11): p. 1485-1487.

3. Herndon, J.M., *Terracentric nuclear fission georeactor: background, basis, feasibility, structure, evidence and geophysical implications.* Current Science, 2014. 106(4): p. 528-541.

4. Elsasser, W.M., *On the origin of the Earth's magnetic field.* Phys. Rev., 1939. 55: p. 489-498.

5. Elsasser, W.M., *Induction effects in terrestrial magnetism.* Phys. Rev., 1946. 69: p. 106-116.

6. Elsasser, W.M., *The Earth's interior and geomagnetism.* Revs. Mod. Phys., 1950. 22: p. 1-35.

7. Herndon, J.M., *Solar System processes underlying planetary formation, geodynamics, and the georeactor.* Earth, Moon, and Planets, 2006. 99(1): p. 53-99.

8. Herndon, J.M., *Nature of planetary matter and magnetic field generation in the solar system.* Current Science, 2009. 96(8): p. 1033-1039.

9. Herndon, J.M., *Geodynamic Basis of Heat Transport in the Earth.* Current Science, 2011. 101(11): p. 1440-1450.

10. Urey, H.C., *The Planets*1952, New Haven: Yale University Press.

11. Lord_Rayleigh, *On convection currents in a horizontal layer of fluid where the higher temperature is on the under side.* Phil. Mag., 1916. 32: p. 529-546.

12. Jackson, A., A. Yonkers, and M. Walker, *Four cenruries of geomagnetic secular variation from historical records. .* Phil. Trans. Roy. Soc. Lond., 2000. 358: p. 990.

13. Bogue, S.W., *Very rapid geomagnetic field change recorded by the partial remagnetization of a lava flow* Geophys. Res. Lett., 2010. 37: p. doi: 10.1029/2010GL044286.

14. Coe, R.S. and M. Prevot, *Evidence suggesting extremely rapid field variation during a geomagnetic reversal.* Earth Planet. Sci. Lett., 1989. 92: p. 192-198.

15. Sagnotti, L., et al., *Extremely rapid directional change during Matuyama-Brunhes geomagnetic polarity reversal.* Geophys. J. Int., 2014. 199: p. 1110-1124.

16. Herndon, J.M., *Nuclear georeactor origin of oceanic basalt $^3He/^4He$, evidence, and implications.* Proceedings of the National Academy of Sciences USA, 2003. 100(6): p. 3047-3050.

17. Herndon, J.M., *New indivisible planetary science paradigm.* Current Science, 2013. 105(4): p. 450-460.

18. Kent, D.V. and F.M. Gradstein, *A Cretacious and Jurassic geochronology.* Bull. Geol. Soc. Am., 1985. 96(11): p. 1419.

19. Cande, S.C. and D.V. Kent, *Revised calibration of the geomagnetic polarity timescale for the Late Cretaceous and Cenozoic.* J. Geophys. Res., 1995. 100: p. 6093.

20. Newitt, L.R., A. Chulliat, and J. Orgeval, -J., *Location of the North Magnetic Pole in April 2007.* Earth Planet. Space, 2009. 61: p. 703-710.

21. Mandea, M. and E. Dormy, *Asymmetric behavior of magnetic dip poles.* Earth Planet. Space, 2003. 55: p. 153-157.

7. Petroleum and Natural Gas Deposits

Where oil is first found is in the minds of men. – *Often cited remark by Wallace Pratt (1885-1981)*

Imagine this hypothetical situation: The year is 1979 and I discover a gold doubloon on a nearby beach. I publish the discovery believing that there will be a stampede of seekers looking to find more. As time passes, no one appears so I return to the beach and find one gold piece after another, each time publishing the discovery and each time finding I remain the lone seeker. Eventually, I have discovered more than any beachcomber ever. Could this be anything else but some absurd fantasy from Rod Serling's "Twilight Zone"? No, this is reality, if one substitutes important scientific discoveries for "gold doubloons" and understands the beach in question as that portion of nature combed by the geosciences. Absurd, oh yes, more absurd and institutionally corrupt than one might imagine. But that is indeed, I wager, reality, with profound implications for humanity.

For thirty five years I have chased one "doubloon" after another without any competition and without any substantive challenges as I've described in the previous chapters. Not convinced? Consider next the fundamentally new implications on the origination of oil and natural deposits that follow from Earth's early formation as a Jupiter-like gas giant as codified in Whole-Earth Decompression Dynamics [1].

Surface observations of oil and gas seeps, the traditional technique for exploration of subsurface petroleum and natural gas deposits, is still currently in use. Surface observations, however, are now augmented by a host of highly sophisticated technologies to detect and determine the extent of those deposits, including surface geochemical investigations and subsurface geophysical investigations, such as seismic, electrical conductivity, and gravity surveys. With oil companies' vast financial resources being brought to bear, and with many highly trained, competent specialists employed in those investigations, how, one might ask, can a lone scientist hope to make a significant contribution? What remains to be discovered?

Long experience has taught me the following about the progress of science. Communities of scientists typically work within a framework of common understanding, rarely questioning the scientific basis of that understanding. But then along comes a scientist with a fundamentally new insight that calls into question the framework of common understanding and sheds light on a new more-correct basis of understanding; a new paradigm. Many, even the most objective individuals, often find it difficult to re-think previously learned concepts in terms of the new paradigm. Their heart isn't in it, nor their heads. But those who are successful at entertaining radically new ideas may thoroughly benefit by making new discoveries and advancing the common understanding.

For nearly half a century, the paradigms of plate tectonics and planetesimal theory of Earth formation dominated geological thinking. Virtually all geological observations, including the origin of petroleum and natural gas deposits, continue to be described in terms of these two flawed paradigms.

The essential geological elements and processes needed for petroleum and natural gas to accumulate are source rock, reservoir, seal and overburden rock, the latter being to facilitate the burial of the others. The sequential processes involved include trap formation and the generation, migration, and accumulation of petroleum [2]. Foreland basins are low-lying regions, adjacent and parallel to mountain belts that are thought to accumulate partly biogenic sediments which are slowly converted into petroleum and natural gas.

Within the framework of plate tectonics, foreland basins are assumed to have formed as the result of tectonic plate collisions. For example, the West Siberian Basin, host to some of the world's greatest petroleum and natural gas deposits, is assumed to have formed with the Ural mountains which are assumed to have thrust up as the European plate collided with the Asian plate during the formation of Pangaea. But the underpinning of plate tectonics, mantle convection, remains physically impossible, as discussed in Chapter 2. Whole-Earth Decompression Dynamics (WEDD), by contrast, the consequence of Earth's early formation as a Jupiter-like gas giant, affords more precise explanations for myriad observations previously attributed to plate tectonics, including the origin of mountains characterized by folding (see Chapter 3). WEDD also leads to new insights on petroleum and natural gas deposits [1, 3].

Following Earth's gas giant phase, after the primordial gases and ices were stripped away from the rocky, compressed kernel, the Archean Eon (4 – 2.5 billion years ago) began with Earth thus compressed. Its surface, devoid of ocean basins, consisted of a closed, contiguous shell of continental rock. This is the one great supercontinent that I call Ottland; present day continents are derived from this unitary supercontinent by successive decompression-driven, fragmentation.

Eventually, internal pressures became sufficiently great that Ottland started to crack [4-8]. Earth's decompression, driven primarily by the stored energy of protoplanetary compression, is augmented by georeactor nuclear fission and radioactive decay energy [9-11]. During decompression Earth's surface increases in diameter. The results are twofold:

- **To provide increased surface area, cracks form, and cracks are subsequently in-filled; and,**
- **To accommodate decompression caused changes in surface curvature, the "extra" surface area, constrained within a continental perimeter, buckles, breaks, and falls over producing fold-mountains, which better conform to the more-decompressed Earth curvature [12].**

In plate tectonics terminology, "rift" refers to the interface of two plates that are beginning to pull apart. In Whole-Earth Decompression Dynamics, by contrast, "rift" refers to the beginning of the formation of a decompression crack. Rifting is responsible for the progressive fragmentation of the continental surface crust into continents. Globally, virtually all major geological activity related to petroleum and natural gas deposits is the consequence of new surface area formation to accommodate decompression-increased planetary volume [1].

Rifting begins with the formation of a decompression crack. Over time, the crack widens, forming a rift-valley or basin. Volcanic eruptions may subsequently occur, depending mainly upon available heat. The rift-basin thus formed becomes an ideal environment for the development of geological strata frequently associated with petroleum and natural gas deposits, such as sandstone layers, and can remain a part of the continental margins even after ocean floor formation. Such ancient rift-basin formations, commonly referred to as sedimentary basins in the oil industry, are the targets for petroleum and natural gas exploration worldwide.

Virtually all petroleum deposits are connected in some way to, or are the consequence of, rifting, sometimes called extension. Continent fragmentation, both successful and failed, begins with rifting. Observations of rifting that is currently taking place at the Afar triangle in northeastern Ethiopia, as well as observations of the consequences of rifting throughout the East African Rift System can help shed light on the nature of petroleum-deposit-related rifting that has occurred elsewhere. (Figure 7.1)

The Afar triangle is the triple junction where the Red Sea rift, the Carlsberg Ridge of the Indian Ocean, and the East African Rift System meet. Seismic tomographic imaging beneath that region shows a very large, low-velocity zone extending to the base of the lower mantle, and referred to as a "superplume" [13, 14]. Because mantle convection is physically-impossible [15], the high $^3He/^4He$ ratios, $R_A>10$, *i.e.*, relative to those measured of air, determined in Afar volcanic basalt [16], indicate deep-mantle heat channeling, which allows the highly mobile, inert, georeactor-produced helium to migrate upward toward regions of progressively lower density.

The rifting-related processes observed at Afar and along the East African Rift System provide all of the crucial components for petroleum deposit formation. Rifting causes the formation of deep basins, as evidenced, for example, by Lake Tanganyika, depth 1.4 km, the second deepest lake in the world, and by Lake Nyasa, depth 0.7 km, the fourth deepest lake, both of which occur as part of the East African Rift System. The observed uplifting caused by swelling from below [17] makes surface land susceptible to erosion, thus providing great amounts of sedimentary material for reservoir rock in-filling of basins. Volcanic-derived sedimentary material may be richly mineralized, assuring strong phytoplankton blooms. Uplifting can sequester sea-flooded lands leading to the formation of marine salt (halite) deposits through desiccation, and can also lead to dome formation.

Afar swelling is important in providing sediments for reservoir sands, delivered by the Congo, Niger and Nile rivers, to continent-edge basins, which were formed by rifting during earlier continent fragmenting and which are Africa's main petroleum provinces. All of the components of reservoir, source and seal are related to the influx of sediments [18]. Local rifting can provide all those components, as well as basin formation. A report published in June 2, 2009 by *East African Business Week* notes that, based upon recent test-well results, a senior official of the U. S. Department of Energy indicated that in Uganda, at the Lake Albert basin of the East African Rift System, the oil reserves could be as much as those of Persian Gulf countries.

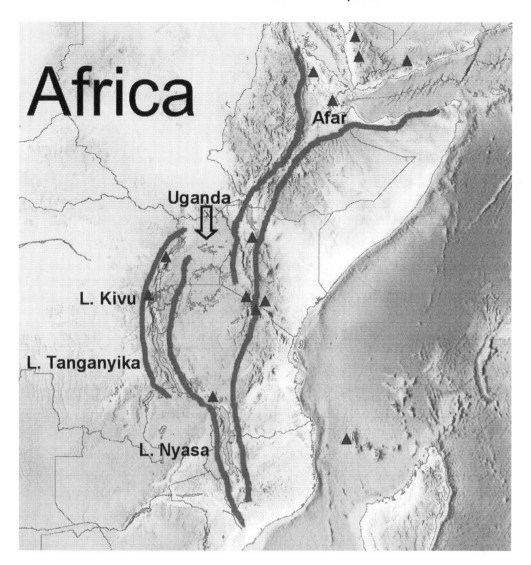

Figure 7.1 Northeastern portion of Africa showing the major rifts comprising the East African Rift System. Active volcanoes are indicated by triangles, oil discoveries by pluses.

Petroleum and natural gas are frequently referred to as "fossil fuels," a designation related to the idea that that these substances formed from buried biologically derived ("biogenic") organic matter. Considerable debate has arisen since Nikolai Kudryavtsev [19] originated what has become the modern Russian-Ukrainian theory of abiotic petroleum, which was widely brought to popular attention in the West by Thomas Gold [20, 21]. Kudryavtsev forcefully argued that no petroleum resembling the composition of natural crude oil had been made from plant material in the laboratory. Moreover, he cited examples of petroleum being found in crystalline and metamorphic basement formations and in their overlying sediments and noted instances of large-scale methane liberation associated with volcanic eruptions. Although still a controversial idea, there has been serious discussion of the subject [22]. Ongoing experimentation further supports the feasibility of hydrocarbon formation under deep-Earth conditions, even in the absence of primary hydrocarbons [23]. Ultimately, however, the prognosis for vast potential resources of abiotic mantle and deep-crust natural gas and petroleum depends critically upon the nature and circumstances of Earth formation.

Earth initially having formed as a Jupiter-like planet is consistent with observations of close-to-star gas giants in other planetary systems [24]. One important point to be made is that the evidence points to the entirety of Earth formation having taken place in intimate association with primordial gases, which includes about 1.3 Earth-masses of methane, the main component of natural gas and a potential progenitor for abiotic hydrocarbons. WEDD, compared to the previous, planetesimal Earth-formation concept, gives a much greater likelihood that carbon compounds were trapped within the Earth during protoplanetary formation. If WEDD is correct, the prognosis is for a deep-Earth methane reservoir, and for a carbon source to abiotically make petroleum [3].

So, what of this possibility of abiotic petroleum? Although there is not yet an unambiguous way to ascertain the extent to which petroleum might be of abiotic origin, the possibility should not be dismissed. Why? Because rifting opens deep cracks which might facilitate contact with methane and perhaps other mantle-derived hydrocarbons. Interestingly, in 1980 Gold and Soter [25] stated that Lake Kivu, part of the East African Rift System, contains 50 million tons of dissolved methane for which there is no adequate microbial source.

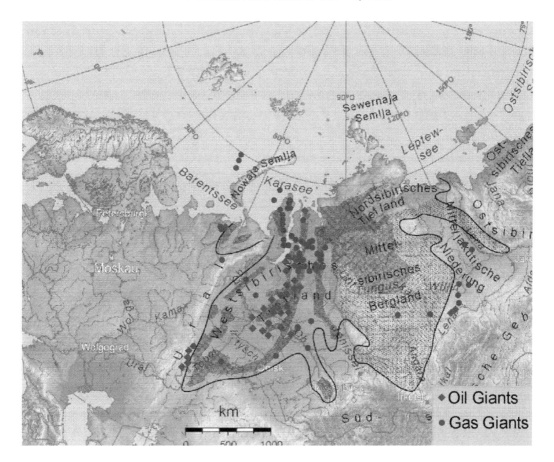

Figure 7.2 Map, courtesy of Jo Weber, showing the extent of the Siberian Traps as based upon estimates derived from [37]. Circles show major gas fields; diamonds show major oil fields. Data from [29].

Consider again Russia's West Siberian Basin, host to some of the world's greatest petroleum and natural gas deposits. But this time consider it from the standpoint of Whole-Earth Decompression Dynamics: The region located between the Ural Mountains and the Siberian Platform, including the West Siberian Basin, underwent extensive rifting about 500-250 million years ago, leading to rift-basin formations. These formations developed geological strata on a grand scale, extremely conducive to trapping petroleum and natural gas [26]. About 250 million years ago massive basalt floods spewed forth for about one million years, blanketing the area with perhaps more than 2,000 km3 of basalt, said basalt containing helium characterized by high ^{3}He/^{4}He ratios [27, 28]. This massive basalt blanket is referred to as the Siberian Traps. Rifting appears to underlie the formation of rift-basins and later the massive eruption of basalt. Evidence indicates that rifting continued after basalt outpouring [26]. The georeactor-signature, high ^{3}He/^{4}He ratios measured in the basalt, suggests that heat was channeled from the Earth's core [15]. Today, that area is known to contain some of the most extensive petroleum and natural gas deposits in the world.

Figure 7.2 is a map showing the extent of the Siberian Traps. I added to that map the sites of major oil and gas wells [29]. The overwhelming proportion of indicated gas and oil fields contained within the Siberian Traps region points to a causal relationship: Decompression-driven rifting, especially when associated with georeactor-produced heat channeled from the Earth's core, is the principal basis for the origination of petroleum and natural gas deposits [1]. The geological structures formed as a consequence of WEDD are ideally suited for petroleum and natural gas, biotic and/or abiotic, as well as for the formation of coal deposits, which to my knowledge are exclusively biogenic.

About 1,100 million years ago, rifting occurred on a grand scale in the middle of what is now the North American continent. The Mid-Continental Rift System or Keweenawan Rift, as it is called, began to split the surface for a length of about 2,000 km from central Kansas to Lake Superior [30]. (Figure 7.3) In a manner similar to Afar and the East African Rift System, a hotspot presumably caused uplift that produced a large dome structure. Massive lava flows erupted from the central axis of the Keweenawan Rift. Thick sedimentary deposits, including a massive oil-shale layer, were emplaced. Rifting ceased before it split the continental mass and formed a new ocean basin, just as was the case with the Siberian Traps.

Suppose that the rifting which gave rise to the Siberian Traps had not failed, but had split the continent and opened a new ocean basin. For such a hypothetical situation, one might expect subsequently to discover the oil and gas deposits along the new continental margins. Although this example is hypothetical, the principle involved may explain generally why oil and gas deposits occur along continental margins; their emplacement began before the rift succeeded in splitting the respective continent and opening a new ocean basin.

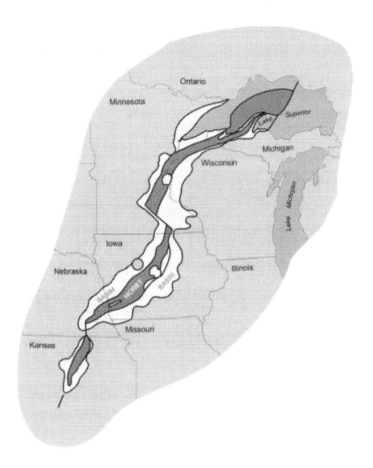

Figure 7.3 Map of Mid-Continent Rift System, courtesy of Raymond R. Anderson and the Iowa Department of Natural Resources.

India can be understood geologically as a continent in the process of fragmenting, the specifics of which are described in detail by Sheth [31]. (Figure 7.4) Note the numerous rifts indicated in that schematic representation of India.

Sheth [31] set forth compelling evidence that India's Deccan Traps originated by rifting. The consequence of rifting is the formation of rift valleys and rift basins, which over geological time may develop into sedimentary basins with geological strata favorable to the entrapment of petroleum and natural gas. Like the Siberian Traps, the Deccan Traps are the consequence of massive flood basalt eruptions about 65 million years ago which blanketed the underlying geological features. Also, like the Siberian Traps, the high $^3He/^4He$ ratios observed in Deccan-basalt is indicative of the heat having been channeled from the Earth's core [32].

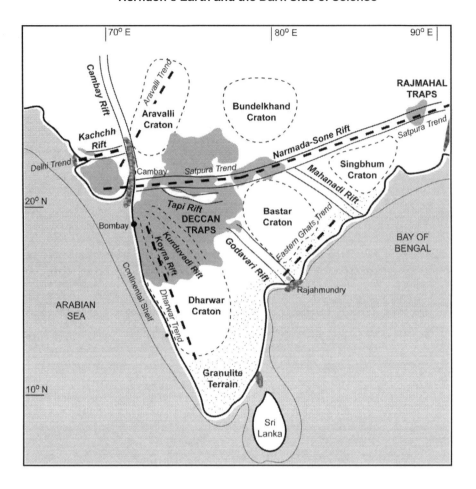

Figure 7.4 Schematic representation of India, adapted with permission from Sheth [31] showing rift zones crossing peninsula India and showing the present outcrop areas of the Deccan and Rajmahal flood basalts (shaded). Areas where significant rift-related petroleum and natural gas discoveries have been made are indicated by dark texture. Not to scale.

From the standpoint of the reasoning developed in the scientific article upon which this chapter is based [1], from the generality of the considerations involved, and from the experience garnered from the larger and older Siberian Traps, I predicted that, as in the Siberian Traps example, the region beneath India's Deccan Traps may well host potentially great petroleum and natural gas deposits. And, indeed the evidence to date seems to support that assertion.

An article in *The Economic Times* (1 May 2008), announced the discovery by ONGC of petroleum and natural gas beneath the Deccan Traps. Moreover, there has been a concerted and successful effort by scientists to detect hydrocarbon traces on the surface that points to the existence of hydrocarbons beneath the Deccan Traps [33-35]. Moreover, a substantial body of evidence in support of the "prospectivity" of the Deccan Traps was presented at a 2013 conference sponsored by the Society of Petroleum Geophysicists, India [36].

Generally, continent fragmenting, as the result of Whole-Earth Decompression Dynamics, augmented by georeactor heat channeled from beneath, whether ongoing, completed or arrested, seems to lead to rift basin formation, to surface uplift and doming, and to the formation of voluminous volcanic deposits. The geological consequences of this can provide all of the crucial components for petroleum and natural gas deposits: basin, reservoir, source and seal. Not surprisingly, many oil and gas exploration activities are focused along continental margins where fragmenting and presumably petroleum deposit formation has taken place.

Chapter 7 References

1.   Herndon, J.M., *Impact of recent discoveries on petroleum and natural gas exploration: Emphasis on India*. Current Science, 2010. 98(6): p. 772-779.

2.   Al-Hajeri, M.M., et al., *Basin and petroleum system modeling*. Oilfield Review, 2009. 21(2): p. 14-29.

3.   Herndon, J.M., *Enhanced prognosis for abiotic natural gas and petroleum resources*. Current Science, 2006. 91(5): p. 596-598.

4.   Herndon, J.M., *Whole-Earth decompression dynamics*. Current Science, 2005. 89(10): p. 1937-1941.

5.   Herndon, J.M., *Solar System processes underlying planetary formation, geodynamics, and the georeactor*. Earth, Moon, and Planets, 2006. 99(1): p. 53-99.

6.   Herndon, J.M., *Energy for geodynamics: Mantle decompression thermal tsunami*. Current Science, 2006. 90(12): p. 1605-1606.

7.   Herndon, J.M., *Nature of planetary matter and magnetic field generation in the solar system*. Current Science, 2009. 96(8): p. 1033-1039.

8.   Herndon, J.M., *New indivisible planetary science paradigm*. Current Science, 2013. 105(4): p. 450-460.

9.   Herndon, J.M., *Indivisible Earth: Consequences of Earth's Early Formation as a Jupiter-Like Gas Giant*, L. Margulis, Editor 2012, Thinker Media, Inc.

10.   Herndon, J.M., *Beyond Plate Tectonics: Consequence of Earth's Early Formation as a Jupiter-Like Gas Giant*, 2012, Thinker Media, Inc.

11.   Herndon, J.M., *Terracentric nuclear fission georeactor: background, basis, feasibility, structure, evidence and geophysical implications.* Current Science, 2014. 106(4): p. 528-541.

12.   Herndon, J.M., *Origin of mountains and primary initiation of submarine canyons: the consequences of Earth's early formation as a Jupiter-like gas giant.* Current Science, 2012. 102(10): p. 1370-1372.

13.   Zhao, D., *Seismic structure and origin of hotspots and mantle plumes.* Earth Planet. Sci. Lett., 2001. 192: p. 251-265.

14.   Ni, S. and e. al., *Sharp sides to the African superplume.* Sci, 2002. 296: p. 1850-1852.

15.   Herndon, J.M., *Geodynamic Basis of Heat Transport in the Earth.* Current Science, 2011. 101(11): p. 1440-1450.

16.   Marty, B. and e. al., *He, Ar, Nd and Pb isotopes in volcanic rocks from Afar.* Geochem. J., 1993. 27: p. 219-228.

17.   Almond, D.C., *Geological evolution of the Afro-Arabian dome.* Tectonophysics, 1986. 331: p. 302-333.

18.   Burke, K., D.S. MacGregor, and N.R. Cameron, *Africa's petroleum systems: Four tectonic "Aces" in the past 600 million years. In: MacGregor, D. S. and Cameron, N. R. (eds) Petroleum Geology of Africa: New Themes and Developing Technologies, Special Publication No. 207. Geological Society, pp 21-60.* 2003.

19.   Kudryavtsev, N., Petroleum Econ. (Neftianoye Khozyaistvo), 1951. 9: p. 17-29.

20.   Gold, T., *The origin of natural gas and petroleum, and the prognosis for future supplies.* Ann. Rev. Energy, 1985. 10: p. 53-77.

21.   Gold, T., *Deep Hot Biosphere*2001, New York: Copernicus Books. 243.

22.   Paropkari, A.L., *Abiogenic origin of petroleum hydrocarbons: Need to rethink exploration strategies.* Curr. Sci., 2008. 95(8): p. 1018-1020.

23.   Kenney, J.F., et al., *The evolutiom of multicomponent systems at high pressures: VI The thermodynamic stability of the hydrogen-carbon system: The genesis of hydrocarbons and the origin of petroleum.* Proc. Nat. Acad. Sci. USA, 2002. 99(17): p. 10976-70981.

24.   Seager, S. and D. Deming, *Exoplanet Atmospheres.* Ann. Rev. Astron. Astrophys., 2010. 48: p. 631-672.

25.   Reichow, M.K., et al., *$^{40}Ar/^{39}Ar$ dates from the west siberian basin: Siberian flood basalt province doubled.* Sci., 2002. 296: p. 1846-1849.

26.   Basu, A.R., et al., *High-^{3}He plume origin and temporal-spacial evolution of the Siberian flood basalts.* Sci., 1995. 269: p. 882-825.

27.   Reichow, M.K., Pringle, M.S., Al'Muk'hamedov, A.I., Allen, M.B., Andreichev, V.L., et al., *The timing and extent of the eruption of the Siberian traps large igneous province: Implications for end-Permian environmental crisis.* Earth Planet. Sci. Lett., 2009. 277: p. 9-20.

28.   Horn, M.K., *Giant Fields 1869-2003*, in *Giant Oil and Gas Fields of the Decade, 1990-1999*, M.K. Halbouty, Editor 2003, AAPG: Houston.

29.   Van Schmus, W.R. and W.J. Hinze, *The midcontinent rift system.* Ann. Rev. Earth Planet. Sci., 1985. 13: p. 345-383.

30.   Sheth, H.C., *From Deccan to Reunion: no trace of a mantle plume*, in *Geological Society of America Special Paper 388*, G.R. Foulger, et al., Editors. 2005, Geological Society of America. p. 477-501.

31.   Basu, A.R., et al., *Early and late alkali igneous pulses and a high-^{3}He plume origin for the Deccan flood basalts.* Sci., 1993. 261: p. 902-906.

32.   Dayal, A.M., D.J. Patil, and S.V. Raju, *Search of hydrocarbons in Mesozoic sediments below the Deccan basalt, India.* Geochim. Cosmochim. Acta, 2008. A204: p. 44.

33.   Dayal, A.M. and e. al., *Geomicrobial prospecting for hydrocarbon research in Deccan Syneclise, India.* Geochim. Cosmochim. Acta, 2009. A270: p. 7.

34.   Rasheen, M.A. and e. al., *Geo-microbial and light gaseous hydrocarbon anomalies in near surface surface soils of Deccan Syneclise Basin, India: Implications to hydrocarbon resource potential.* J. Petrol. Sci. Engr., 2012. 84-85: p. 33-41.

35.   Jamkhindikar, A., M. Jain, and S.N. Mohanty. *Hydrocarbon Prospectivity of Deccan Trap in Northern Cambay Basin* in *10th Biennial International Conference & Exposition 2013.* 2013. Kochi, India.

36.   Masaitis, V.L., *Permian and Triassic volcanism of Siberia.* Zapiski VMO, 1983. part CXII(4): p. 412-425.

8. Solar System Briefly

I know that most men, including those at ease with problems of the greatest complexity, can seldom accept even the simplest and most obvious truth if it be such as would oblige them to admit the falsity of conclusions which they have woven, thread by thread, into the fabric of their lives.— *Tolstoy*

In the natural sciences it is difficult to prove the existence of some event or process, some might even say impossible. So, how can one ascertain what really happened in nature? How can one separate truth from misunderstanding/deceit? From my experience it is this: One begins from the foundation of knowledge, securely anchored to the properties of matter and/or radiation. Then, one progresses logically step-by-step to reveal causal relationships that lead to advances. I call that method a logical progression of understanding [1]. At all times one must be absolutely truthful; deceit leads nowhere but to confusion. Debate and discussion can lead to new insights and help to reveal errors, which, if present, should be corrected promptly. When all that is done, the new understanding "makes sense," which is another way of saying that the components are related logically and causally, like the scenes of a really good movie. That methodology should be clear from the previous chapters in which I describe the new indivisible geoscience paradigm, Whole-Earth Decompression Dynamics, wherein practically all geophysical, geochemical, and geological considerations follow logically from Earth's protoplanetary origin and early formation as a Jupiter-like gas giant. My new indivisible geoscience paradigm stands in striking contrast to the so-called "standard model of solar system formation" that is, I regret to say, an accumulation of *ad hoc* assumptions stuck together within a framework of physical impossibilities, rather like the scenes of a very bad movie – a movie on a specialized topic whose under-budgeting or lack of care did not allow consultation with people who knew the topic before being rushed into theaters.

As Earth is but one planet in our Solar System, it should come as no surprise that I applied a similar methodology to the Solar System as a whole, with all its planets, and in 2013 published a major scientific article, "New

Indivisible Planetary Science Paradigm" in *Current Science* [2]. (See Appendix C for a copy of that scientific article, which contains considerably greater detail than the present chapter.)

Recall the chondrite meteorites, debris from outer space whose compositions are relatively unchanged from the makeup of matter in the outer part of the Sun, except for volatile elements or those that form volatile compounds. Although being similar to one another in elemental composition, for many elements within a factor two, the chondrites form three distinct groups with different mineral compositions and three distinct states of oxidation. The rare, highly-reduced enstatite chondrites, *e.g.*, the Abee meteorite, may be understood as having formed from matter that originally condensed, and was isolated from primordial gas at high-temperatures and high-pressures – such as one would expect to prevail in the inner Solar System during the time of the Sun's formation. The rare and highly oxidized carbonaceous chondrites, such as the Orgueil meteorite, may be understood as having formed from matter that originally condensed at very low-temperatures at very low-pressures that might have prevailed in the outer regions of the Solar System. But what, you might ask, about the ordinary chondrites? Ordinary chondrites that are the most abundant meteorites, and are the ones that inspired the mistaken basis of deep-Earth composition?

For more than half a century, I suggest, the mistaken belief that Earth resembles an ordinary chondrite has adversely impacted geoscience without the origin of ordinary chondrites ever having been known [3]. Only five chemical elements comprise 95% of the mass of each chondrite: Iron (Fe), silicon (Si), and magnesium (Mg) plus the two elements that combine with them, oxygen (O) and sulfur (S). For decades, the abundances of major elements in chondrites have been expressed in the literature as ratios, usually relative to silicon (E_i/Si) and occasionally relative to magnesium (E_i/Mg). By expressing Fe-Mg-Si elemental abundances as atom (molar) ratios relative to iron (E_i/Fe), as shown in Figure 8.1, I discovered a fundamental relationship bears on the origin of ordinary chondrite matter [4].

The relationship I discovered admits the possibility of ordinary chondrites having been derived from mixtures of two components, representative of the other two types of matter, mixtures of a relatively undifferentiated carbonaceous-chondrite-like *primitive* component (point A) and a partially differentiated enstatite-chondrite-like *planetary* component (point B). "Differentiated" here means that the element composition has been changed somewhat from primordial "solar" composition. The primitive fraction and the planetary fraction may be determined for each ordinary chondrite, and then may be used to better characterize the compositions of the two reservoirs from which each ordinary formed.

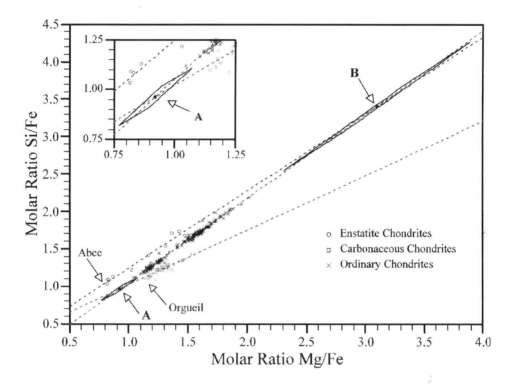

Figure 8.1 Atom (molar) ratios of Mg/Fe vs. Si/Fe from analytical data on 10 enstatite chondrites, 39 carbonaceous chondrites, and 157 ordinary chondrites. The well-defined lines are evident only when normalized to Fe, not to Si or Mg. The ordinary chondrite points scatter about a line that intersects the other two lines. Points on the ordinary chondrite line can be represented by mixtures of the two intersecting compositions, point A: *primitive*, and point B: *planetary*. The 95% confidence intervals are shown by solid lines only near points of intersection. For more detail, see [4].

The relationship shown in Figure 8.1 is a special circumstance for those three major elements of chondrites. It is possible to estimate the compositions of point A and point B on Figure 8.1 from primitive fraction or planetary fraction correlations with ordinary chondrite chemical composition data. Significantly, the planetary component (point B) is partially, not fully, differentiated; it still consists of about one-third of its original amount of iron alloy along with the silicates. Refractory (less-volatile) siderophile (iron-loving) elements, such as iridium (Ir) and osmium (Os), however, are depleted seven-fold in the planetary component relative to the more volatile siderophile elements, nickel (Ni), cobalt (Co), and gold (Au). This is evidence of heterogeneous planetary formation – suggesting that the first iron to condense took with it the refractory siderophile elements.

Try to visualize this: Imagine a giant gaseous protoplanet with high-pressures and high-temperatures at its center with liquid iron metal just starting to condense and to in-fall to the center. The first molten iron to condense will dissolve and/or scavenge the refractory, high temperature-condensing metals, such as iridium, and remove those by in-falling to the center where they accumulate as the planetary core. Suppose during that process, before the planet had fully formed, the violent T-Tauri solar winds stripped away matter and transported it to the region between Mars and Jupiter. That is what I think happened to Mercury's incompletely formed protoplanet, which left Mercury with a disproportionately smaller mantle than if the violence had not occurred.

So, this is the origin of the planetary component: For the reasons described in my scientific publication [4], the planetary component, I posited, was the partially differentiated matter stripped from Mercury's protoplanet by the T-Tauri super-intense solar winds.

When one understands a problem and its solution, it is possible to write both the problem and its solution on the back of an envelope. The back of my envelope looks like this: The planets formed as gas giants, literally raining out from within giant gaseous protoplanets at high pressures. The super-intense T-Tauri outbursts associated with the ignition of the Sun stripped the gases from the inner planets and stripped away some of the matter from not-yet completed Mercury. That stripped away matter fused with low-pressure, low-temperature condensate, similar in oxidation state to the Orgueil carbonaceous chondrite, to form the parent matter of ordinary chondrites and most asteroids in the region of space between Mars and Jupiter. Viewed in that context, ordinary chondrite matter may be considered as a veneer that fell onto the outer portion of Earth, and to a greater degree, onto Mars.

One of the seemingly inexplicable mysteries of modern planetary science is why the terrestrial planets display such diverse surface dynamics even though those planets are more-or-less chondritic in composition? The solution to that mystery, I posit, is that the terrestrial planets experienced different degrees of compression by primordial gases and ices: Mercury never experienced the Jupiter-like gas/ice shell that Earth experienced; Mars may have experienced little or none. The degree of compression experienced by Venus cannot yet be determined. The surface dynamics of Earth are the consequence of our planet's early formation as a Jupiter-like gas giant. The weight of about 300 Earth-masses of gas/ice compressed the rocky part; Earth's subsequent decompression following removal of its massive gas/ice shell drives and causes its surface geology:

- **The repeated splitting of continents and the concomitant formation of ocean basins to increase surface area, and;**
- **The formation of mountains characterized by folding to compensate for surface curvature changes**

Earth's decompression also emplaces heat at the base of the crust, which, I posit, is the origin of the geothermal gradient.

As one progresses downward through the crust, by drilling or mining, the temperature increases. This is referred to as the geothermal gradient. On Earth, the geothermal gradient limits the downward capillary-action percolation of water. Mars never formed as a Jupiter-like gas giant, at least not to the extent Earth did, so Mars was not very compressed and consequently did not decompress. Mars' absence of significant decompression means that that planet essentially lacks a "geothermal gradient" which means that water can settle deeper into the planet. In other words, Mars has potentially a greater subsurface reservoir capacity than perhaps realized.

I have suggested that only three processes, operant during the formation of the Solar System, are primarily responsible for the diversity of matter in the Solar System and are directly responsible for planetary internal compositions and structures [3]. These are:

- **High-pressure, high-temperature condensation from primordial matter associated with planetary-formation by condensing from the interiors of giant-gaseous protoplanets, in-falling and accumulating at the planetary centers;**

- Low-pressure, low-temperature condensation from primordial matter in the remote reaches of the Solar System or in the interstellar medium; and,
- Stripping of the primordial volatile components from the inner portion of the Solar System by T-Tauri outbursts, presumably during the thermonuclear ignition of the Sun.

The commonality of these three processes provides a way to understand the nature of matter within extraterrestrial planetary interiors and the nuclear fission that, as with our own proposed georeactor, provides their internal heat and generates their magnetic fields [2, 5].

Figure 8.2 The four giant gaseous planets, Jupiter, Saturn, Uranus and Neptune. Note the pronounced turbulence, driven by internal heat, in all but Uranus.

Prior to 1969, scientists thought that planets do not produce energy, except small amounts from radioactive decay; planets just receive energy from the Sun and then radiate it back into space. Then, astronomers observed that Jupiter, Saturn and Neptune radiate into space nearly twice the energy they receive from the Sun. (Figure 8.2) For twenty years the source of that internal energy was a mystery to NASA-funded scientists, who wrongly thought they had considered and eliminated all possibilities. So, what did NASA scientists propose for an explanation of the origin of this anomalous heat? They assumed that the giant planets, even after 4½ billion years are still collapsing and changing gravitational potential energy into kinetic energy. But that explanation did not make sense, at least to me: The giant planets consist mainly of a mixture of hydrogen and helium, both of which are extremely fluid and excellent heat transfer media. As you might expect, another idea presented itself.

One day while shopping for groceries, BANG! Like a bolt of lightning, I suddenly realized that Jupiter has all the ingredients for a natural, planetary-scale nuclear reactor at its center. I did the necessary research and calculations. Then in 1991, I submitted a scientific paper to the German *Naturwissenschaften* demonstrating the feasibility of that energy being produced by natural nuclear fission reactors at the giant planets' centers. I applied Fermi's nuclear reactor theory using the same approach that Paul K. Kuroda had used in 1956 to predict the occurrence of natural nuclear reactors in ancient uranium mines. Recall that although Kuroda was all but ignored by his fellow scientists, nature proved his predictions correct when the fossil remains of natural nuclear reactors were discovered in 1972 in a uranium mine at Oklo, in the Republic of Gabon in Western Africa.

When my paper was accepted for publication [6], I submitted a research proposal to NASA's Planetary Geophysics Program. Paul K. Kuroda accepted my invitation to join in as a co-investigator. Kuroda graciously insisted that his efforts be *pro bono* as he "did not need the money."

The Universities Space Research Association, an association of major institutional recipients of NASA funding, operates the Lunar and Planetary Institute, which operated the Lunar and Planetary Geoscience Review Panel (LPGRP) at the time I submitted the proposal. The LPGRP served NASA by soliciting anonymous "peer reviews" of submitted proposals, then, based upon those "peer reviews," evaluating the proposals in secret session, and then ranking them so as to make it easy for a NASA official to decide which to fund

The LPGRP, composed of a group of principal investigators of NASA grants, funded either through NASA's Planetary Geophysics Program or its Planetary Geology Program, conducted the ranking of all proposals submitted to one or the other of those same two NASA programs. In other words, my proposal was competing for the same limited pool of funds as proposals from the very institutions whose personnel served on the LPGRP. At the time, the chairman of the LPGRP was associated with NASA's Jet Propulsion Laboratory, which is operated by the California Institute of Technology (Caltech), and which consumed more than 40% of the budget of the Planetary Geophysics Program.

Normally, the LPGRP's ranking of proposals is kept secret, but through extraordinary efforts I learned from the U. S. Congress' General Accounting Office (called the Government Accountability Office since 2004) that on technical merit the LPGRP ranked my proposal lowest of the 120 proposals submitted to NASA's Planetary Geophysics Program. In retrospect, one might seriously question the integrity of that ranking, as I later independently performed all that I had proposed and much more, including demonstrating the feasibility of a nuclear fission reactor at the center of Earth, called the georeactor, as the energy source and production mechanism for the Earth's magnetic field [7-9]. I also extended the concept to other planets, as stated above, as well as to their large moons [5]. (Appendix C) Since then, the concept of planetary nuclear fission reactors has received quite thorough vetting in the international scientific community. (Appendix B) So, what was NASA's response?

Now, more than twenty years after the first publication of my planetary nuclear fission papers [6, 7, 10], NASA-supported planetary scientists are, as before, promulgating the idea that the giant planets are still cooling, collapsing and releasing heat [11]; variations on the same old story. What is wrong now, though, is that they are either not aware of the full spectrum of scientific possibilities, or deliberately obfuscating, as their papers never even reference the published, more economical and logically consistent possibility of the existence of planetary nuclear fission reactors.

The tiny planet, Mercury, with a mass of about 5% that of Earth, is an excellent example of planetary circumstances that might be better explained by considering a central nuclear fission reactor. (Figure 8.3) Mercury is also an excellent example of NASA-supported scientists systematically refusing to make any reference to the planetocentric nuclear reactor concept even though they make references to a variety of ideas that are on considerably less firm scientific footing.

Figure 8.3 Planet Mercury, composite image from NASA's MESSENGER mission.

Debate over Mercury's heat source and presumptive magnetic field generation has long existed. In 1974-75, the *Mariner 10* spacecraft flew by Mercury three times and discovered that Mercury has a global magnetic field. The nature of that magnetic field and its energy source has been the subject of much discussion by NASA-supported scientists. Fifteen years before my concept of planetocentric nuclear fission reactors was published, Sean C. Solomon, later the Principal Investigator for NASA's MESSENGER Mission to Mercury, wrote:

A convective dynamo mechanism for Mercury's magnetic field is in apparent conflict with cosmochemical models that do not predict a

substantial source of heat, most probably radiogenic, in Mercury's core. Without such a heat source, the core would solidify within about 1 b.y. [billion year] .... [12].

The September 30, 2011 issue of *Science* magazine was devoted to reports by the Project MESSENGER science team members of that NASA mission to orbit Mercury. The Principal Investigator and members of that team, however, neglected to mention or cite my planetary nuclear fission reactor scientific publications [5-8, 10, 13-16], this despite the fact that the planetary nuclear reactor concept neatly explains the existence of an unanticipated, substantial heat source at Mercury's center. It also explains a mechanism, unanticipated within the standard model, for generating Mercury's magnetic field. Instead of giving these ideas a fair hearing and realizing they might explain the phenomena in question, NASA-funded scientists published a plethora of core-dynamo models based upon arbitrary assumptions [17, 18] and even published a thermoelectric model that requires even further assumptions [19]. There were no references to the peer-reviewed scientific publications that bear on Mercury's origin, internal heat source, mode of magnetic field generation, and low-FeO content of Mercury's surface material that resembles the silicate-rock portion of an enstatite chondrite [4, 5, 7, 8, 10, 13-16, 20-22].

At the frontiers of science, at the interface of the unknown, it may be that one cannot specify which of several possibilities might be correct or whether, perhaps, the correct possibility has not yet been envisioned. In such instances, ethical scientists should and do cite the various suggested possibilities, including ideas other than their own that are published in world-class scientific journals.

Nuclear Moons

Ganymede, one of Jupiter's moons, has a diameter about 8% greater than that of Mercury. (Figure 8.4) Like Mercury, Ganymede has an internally generated magnetic field that seemed to bewilder NASA-funded planetary scientists.

Figure 8.4 Jupiter's moon Ganymede, which has an internally generated magnetic field.

They stated that the interior of the Ganymede would have cooled and solidified in the first 1 to 2 billion years making a convection-driven dynamo impossible [23, 24]. But, whether ignorant of the published literature or steered away from it, the NASA-funded planetary scientists failed to mention the exciting, testable, and partially tested possibility of natural extraterrestrial nuclear fission, and georeactor-like off Earth nuclear fission reactors. Perhaps, if they had, one of them might have realized that such a nuclear reactor is not only a heat source but potentially a self-regulated, convection-driven dynamo for magnetic field production. [5, 8].

Figure 8.5 Jupiter's moon Io, which has an internally generated magnetic field. Note the volcanically scarred surface.

Jupiter's moon, Io, is currently the most volcanically active object in the Solar System [25, 26]. (Figures 8.5 and 8.6) With a diameter that is just 75% that of Mercury, Io is thought to have an internally generated magnetic field [27]. Although at first Io's volcanic eruptions were thought to consist of sulfur, which has a relatively low melting point, later observations revealed much higher temperatures consistent with silicate volcanism [25, 26]. Io's surface heat flow has been estimated at more than 2.5 Watts per square meter, W/m² [28]. This is more than twice the heat calculated to arise from the gravitational push-and-pull (tidal interaction) with Jupiter and its other moons [29]. Without the natural nuclear reactor hypothesis, that is hard to explain.

Figure 8.6 Ongoing volcanic eruption at Tvashtar Catena on Jupiter's moon Io.

Linda T. Elkins-Tanton, whose research relates to molten magma, writes [24]:

> Tidal stresses are not the final word .... Io's power output from volcanism is about 2.5 W/m^2. This value may be more than the energy contributed from Jupiter through tidal heating (some researchers contend that it is more than twice the energy that Jupiter contributes). It is about twice the magnitude of the heating provided by electromagnetic heating from the ion storm around Io, and it also exceeds any heat possible from radioactive decay. Io's heat output, in fact, is more than twice that of Earth's. Tidal heating is still

accepted as the method for creating the heat required for the extravagant quantity of volcanic activity on Io.

The word "accepted" in the last sentence begs the question: Accepted by whom? Let me be blunt for a moment. NASA-funded scientists might pretend that it is their responsibility to accept or not accept science, but that is nonsense. The responsibility of NASA-funded scientists is to try to refute ideas they consider contradictory and, if they are unable to refute them, then they need to acknowledge those ideas. That is the historical path of ethical science in pursuit of truth about the universe for its own sake rather than, let us say, being casualties of institutional inertia. Bluntly said, Elkins-Tanton does a disservice to the public by implying that "accepted" is somehow an acceptable measure of scientific quality, replacing scientific investigation itself. It is not.

So, let me fantasize what Elkins-Tanton might have written in place of the above last sentence? I suggest the following:

It appears that Io contains an energy source not considered by NASA scientists, perhaps a nuclear fission reactor of the type proposed by Herndon [5-8, 10, 13-16].

Trouble at NASA – Open Science versus Closed Bureaucracy

The Solar System does not alter its essential behavior or structure because of what we think about it. The objects in our Solar System, as in the universe, are the consequence of the processes of their formation and the actions of subsequent processes. As a scientist, the best one can do is to try to understand those objects and the processes that brought them to their present states. Although the advances I made may not be "politically correct" within present NASA culture, the underlying science is solid, relevant to precise problems investigated by NASA at taxpayer expense, and published in world-class scientific journals. And I have to wonder whether this is my experience alone. How many other capable scientists has NASA (among other science-funding agencies) driven away by this convoluted, in-crowd behavior? How many scientific advances have been stopped in their tracks, or permanently lost due to bureaucracy, ignorance, institutional malfeasance, and ego? That is impossible to say. I am probably the exception. Rather than going away, I continued to make scientific discoveries and insights at my own expense, while NASA-funded scientists muddle in a flawed but comfortable storyline "protected" from the thrill of comparative analysis to find out, via evidence, research and reasoning, the most likely history, changes, and composition of our Solar System.

Those institutions and individuals who receive NASA funding for scientific investigations cannot challenge the NASA-storyline. Otherwise, in secret reviews, one or more NASA-funded individuals is apt to unwarrantedly berate proposals for funding or papers submitted for publication. "Career-fear" is the operative unscientific principle. It is pervasive in the world of NASA secret reviews. So it is that the majority remain silent and capitulate. As a result of such behavior many may never even realized that my work may be deeply relevant to their own. We need a more curious, critical attitude to science and science education. Many of these NASA-supported scientists I have been complaining about are highly capable professors at prestigious American universities. These are the people who should be teaching students, in words and by their own example, with great enthusiasm, that science is about telling the truth, the full truth. Instead, they are training generations of scientists, by example if not jaded and coded words, never to contradict the "politically correct" stories for fear of ostracism or loss of grants. Science I am afraid has unfortunately become politicized to its own detriment.

And it is actually even worse. NASA has hurt education with grants to science teacher professional organizations for the purpose of further training science teachers. NASA has infected education with grants aimed at training science teachers or, perhaps better said, *indoctrinating* science teachers. Science is all about questioning extant ideas, observations, and theories. But how often do NASA-grant-receiving organizations teach teachers to question NASA's science? Questioning NASA's science is a "politically correct" no-no. There is a widespread perception, real or imagined, that to question NASA's science will be grant-suicide. In this manner NASA has unwittingly and paradoxically become an active part in the dumbing-down of American science education.

I was fortunate to have learned science from Nobel laureate Harold C. Urey (1893–1981), who had learned science from Nobel laureate Niels Bohr (1885–1962). I also learned science from geoscientist Hans Eduard Suess (1909–1993), who learned science from his father, geoscientist Franz Eduard Suess (1867–1942), who learned science from his father, geoscientist Eduard

Suess (1831–1914), the person who coined the term "biosphere." It has been thirty-five years since I was "excommunicated" for challenging the then forty-year-old idea of the inner core's composition. During those years I have unfortunately witnessed the geoscience community of the university-government complex progressively debasing science standards, substituting making models based upon assumptions for discoveries, and misrepresenting the state of knowledge. I have seen it shying away from the full truth, ignoring scientific advances, and failing to cite contradictions to the "politically correct" storyline. There is at work, I regret to say, an ongoing malevolent political agenda, potentially devastating to humanity and to our planet, and it is being driven by a scientific community that cannot be trusted to tell the truth, a scientific community comprised of fund-recipients of DOE, NASA, and NSF grants/contracts, and U. S. Government scientists and administrators, who have proven themselves inept, irresponsible, and incapable of rendering valid and truthful scientific knowledge.

So, how did this start? What made it possible? My decades of unsupported research on geoscience, Solar System composition and concomitant processes have been intellectually exciting but socially frustrating. I have been driven to take my case directly to lay readers like you. My specialty and interests have always been in the hard sciences, and piecing together the puzzle of the fascinating workings of our amazing Solar System. But since I have been blocked, quite unfairly in my opinion as is obvious from the foregoing plaints, from getting a fair hearing, I will now turn from planetary science to social science, in an effort to see what went wrong and what we can do about it. But as it is in my nature to look ever broader and deeper into science, I would be remiss not to share some new insights as to why the multitude of galaxies have the very few characteristic luminous star distributions that are seen throughout the observable universe.

Chapter 8 References

1. Herndon, J.M., *Inseparability of science history and discovery*. History of Geo- and Space Science, 2010. 1: p. 25-41.

2. Herndon, J.M., *New indivisible planetary science paradigm*. Current Science, 2013. 105(4): p. 450-460.

3. Urey, H.C. and H. Craig, *The composition of stone meteorites and the origin of the meteorites*. Geochim. Cosmochim. Acta, 1953. 4: p. 36-82.

4. Herndon, J.M., *Discovery of fundamental mass ratio relationships of whole-rock chondritic major elements: Implications on ordinary chondrite formation and on planet Mercury's composition*. Current Science, 2007. 93(3): p. 394-398.

5. Herndon, J.M., *Nature of planetary matter and magnetic field generation in the solar system*. Current Science, 2009. 96(8): p. 1033-1039.

6. Herndon, J.M., *Nuclear fission reactors as energy sources for the giant outer planets*. Naturwissenschaften, 1992. 79: p. 7-14.

7. Herndon, J.M., *Planetary and protostellar nuclear fission: Implications for planetary change, stellar ignition and dark matter*. Proceedings of the Royal Society of London, 1994. A455: p. 453-461.

8. Herndon, J.M., *Nuclear georeactor generation of the earth's geomagnetic field*. Current Science, 2007. 93(11): p. 1485-1487.

9. Herndon, J.M., *Terracentric nuclear fission georeactor: background, basis, feasibility, structure, evidence and geophysical implications*. Current Science, 2014. 106(4): p. 528-541.

10. Herndon, J.M., *Feasibility of a nuclear fission reactor at the center of the Earth as the energy source for the geomagnetic field*. Journal of Geomagnetism and Geoelectricity, 1993. 45: p. 423-437.

11. Elkins-Tanton, L.T., *Uranus, Neptune, Pluto, and the Outer Solar System*2011, New York, NY: Facts on File, Inc.

12. Solomon, S.C., *Some aspects of core formation in Mercury*. Icarus, 1976. 28(4): p. 509-521.

13. Herndon, J.M., *Sub-structure of the inner core of the earth.* Proceedings of the National Academy of Sciences USA, 1996. 93: p. 646-648.

14. Herndon, J.M., *Nuclear georeactor origin of oceanic basalt $^3He/^4He$, evidence, and implications.* Proceedings of the National Academy of Sciences USA, 2003. 100(6): p. 3047-3050.

15. Herndon, J.M., *Solar System processes underlying planetary formation, geodynamics, and the georeactor.* Earth, Moon, and Planets, 2006. 99(1): p. 53-99.

16. Hollenbach, D.F. and J.M. Herndon, *Deep-earth reactor: nuclear fission, helium, and the geomagnetic field.* Proceedings of the National Academy of Sciences USA, 2001. 98(20): p. 11085-11090.

17. Gómez-Pérez, N. and S.C. Solomon, *Mercury's weak magnetic field: Result of magnetospheric feedback?* Europ. Planet. Sci. Cong., 2010. 5.

18. Stanley, S. and G.A. Glatzmaier, *Dynamo models for planets other than Earth.* Space Sci. Rev., 2010. 152: p. 617-649.

19. Giampieri, G. and A. Balogh, *Mercury's thermoelectric dynamo model revisited.* Planet. Space Sci., 2002. 50: p. 757-762.

20. Herndon, J.M., *Composition of the deep interior of the earth: divergent geophysical development with fundamentally different geophysical implications.* Phys. Earth Plan. Inter, 1998. 105: p. 1-4.

21. Herndon, J.M., *Mercury's protoplanetary mass.* arXiv:astro-ph/0410009 1 Oct 2004, 2004.

22. Herndon, J.M., *Total mass of ordinary chondrite material originally present in the Solar System.* arXiv: astro-ph 0410242, 2004.

23. Johnson, T.V., *A look at the galilean satellites after the Galileo Mission.* Physics Today, 2004: p. 77-83.

24. Elkins-Tanton, L.T., *Jupiter and Saturn* 2006, New York: Infobase Publishing. 220.

25. McEwen, A.S., et al., *Active volcanism on Io as seen by Galileo SSI.* Icarus, 1998. 135(1): p. 181-219.

26. Stransberry, J.A., et al., *Violent silicate volcanism on Io in 1996*. J. Geophys. Res., 1997. 24(20): p. 2455-2458.

27. Kivelson, M.G., et al., *A magnetic signature at Io: Initial report from the Galileo magnetometer*. Sci., 1996. 273: p. 337-340.

28. Veeder, G.J., et al., *Io's heat flow from infrared radiometry: 1983–1993*. J. Geophys. Res., 1994. 99(E8): p. 17095-17162.

29. Moore, W.B., *Tidal heating and convection in Io*. J. Geophys. Res., 2003. 108(E8): p. 5096-5112.

9. Governance and Galaxies

The very word "secrecy" is repugnant in a free and open society; and we are as a people inherently and historically opposed to secret societies, to secret oaths and to secret proceedings. We decided long ago that the dangers of excessive and unwarranted concealment of pertinent facts far outweighed the dangers which are cited to justify it...a wise man once said: "An error doesn't become a mistake until you refuse to correct it."...[T]he Athenian law-maker Solon decreed it a crime for any citizen to shrink from controversy. And that is why our press was protected by the First Amendment – the only business in America specifically protected by the Constitution – not primarily to amuse and entertain...not to simply "give the public what it wants" – but to inform, to arouse, to reflect, to state our dangers and our opportunities... — *John F. Kennedy, to the* American Newspaper Publishers Association, New York City, *April 27, 1961*

As a student I recall being shocked to learn that much of what is published is not true, it is simply someone's model based upon assumptions, rather than a discovery that is securely anchored to the properties and behavior of matter and radiation. I was shocked because I had believed that the purpose of science is to determine the true nature of Earth and universe. I still believed that. What happened? Why did it happen?

Prior to World War II, although money for science at the time was in short supply, scientists maintained a kind of self-discipline. A graduate student working on a Ph.D. degree was expected to make a new discovery to earn that degree, even if it meant starting over after years of work because someone else made the discovery first. Self-discipline was also part of the scientific publication system. At the time, when a scientist wanted to publish a paper, the scientist would send it to the editor of a scholarly journal for publication and generally it would be published. A new, unpublished scientist was required to obtain the endorsement of a published scientist before submitting a manuscript. The concept of "peer review" had not yet been born.

Before World War II, there was very little government funding of science, but that changed because of war-time necessities. In 1951, the U. S. National Science Foundation (NSF) was established to provide support for post-World War II civilian scientific research. The process for administrating the government's science-funding, invented in the early 1950s by NSF, has been adopted, essentially unchanged, by virtually all subsequent U. S. Government science-funding agencies, such as the National Aeronautics and Space Administration (NASA) and the U. S. Department of Energy (DOE).

The problem, I contend, is that the science-funding process that the NSF invented and passed on to other U. S. Government agencies is seriously and fundamentally flawed. Here follow the principal flaws:

•**NSF FLAW #1: Peer Review.** Proposals for scientific funding are generally reviewed by anonymous "peer reviewers." NSF invented the concept of "peer review," wherein a scientist's competitors would re-view and evaluate his/her/their proposal for funding, and the reviewers' identities would be concealed. The idea of using anonymous "peer reviewers" must have seemed like an administrative stroke of genius because the process was adopted by virtually all government science-funding agencies that followed and almost universally by editors of scientific journals. Ironically, the impetus was to make the process fair, allowing one to be candid without hurting a colleague's feelings. But no one seems to have considered the lessons of history with respect to secrecy. Secrecy is certainly necessary in matters of national security and defense. But in civilian science, does secrecy with its concomitant freedom from accountability really encourage truthfulness? If secrecy did in fact lead to greater truthfulness, secrecy would be put to great advantage in the courts. Courts have in fact employed secrecy – during the infamous Spanish Inquisition and in virtually every totalitarian dictatorship – and the result is always the same: Unscrupulous individuals falsely denounce others and corruption abounds. The application of anonymity and freedom from accountability in the "peer review" system gives unfair advantage to those who would unjustly berate a competitor's proposal for obtaining funding for research and for publishing research results. Secrecy and science are inimical. Openness is the byword of curiosity, and the communication upon which real science thrives. Secrecy breeds an opposite, closed culture, convenient for power games and hierarchy but anathema to science. Anonymous "peer review" has become the major science-suppression method of the new science-barbarians. Moreover, even the perception – real or imagined – that some individuals would, protected behind masks of secrecy, do just what other power grabbers and power retainers have done, has had, and continues to have a chilling effect, putting scientists (who weren't warriors to begin with) on the defensive. The result is wide adoption of "politically correct" consensus-approved viewpoints. Individuals benefit, but culture, and the search for truth, suffers. In more open, less institutionalized cultures in which science was spawned, however, there is less fear and more free flow of information. Instead of backbiting and cheap shots from behind fortified parapets, there is open criticism and response. The richness of various viewpoints universally exposed to criticism invigorates the quest for exploration, keeping people questioning and open to better explanations, benefiting the scientific community and the public at large.

By contrast today the peer review system tends to stymie discussion of anything that might be considered a challenge to others' work or to the funding agencies' programs. And that is not what science is about at all. Not surprisingly, there exists today a widespread perception that to challenge scientific results supported by a U. S. Government agency will lead to loss of one's own financial support. Whether true or not, as mentioned, this perception has a chilling effect.

•NSF FLAW #2: Preplanned Results. NSF invented the concept of scientists proposing specific projects for funding, which has led to the trivialization and bureaucratization of science. Why so? The problem is that it is absolutely impossible to say beforehand what one will discover that has never before been discovered, and to say what one will do to discover it. As Sam Houston State University physics professor C. Renée James has chronicled with scholarly verve, a straight-line attempt to do science for economic reasons not only hurts the science but radically decreases the chances for economic breakthroughs. Curiosity-driven science among open, communicating scientists has led from pink Yellowstone slime to genetic fingerprinting, from twitching frog legs to the first battery, from atomic clocks checking relativity in airplanes to GPS, and even from failed sightings of black hole evaporation to WiFi.

The consequence of putting unrealistic limits on the curiosity motor which powers science by insisting on results in advance of the quest that would find them has been the proposing of trivial projects with often non-scientific end-results, such as the widespread practice of making models based upon assumptions, instead of making discoveries. Even in economic terms we might say, nothing ventured, nothing gained. And from a biological perspective we might compare the shortsighted dissuasion of multiple investigable scientific perspectives to an ecosystem that is not diverse enough to be robust. Contrary to such a curiosity- and experiment-nurturing garden, bureaucratic "program managers" in today's sterile gardens decide which projects are suitable for the programs that they themselves design. Moreover, proposal "evaluation" is often a guise for "program managers" and "peer reviewers" to engage in exclusionary and ethically questionable, anti-competitive practices. There is no incentive for scientists to make important discoveries or to challenge existing ideas; quite the contrary.

•**NSF FLAW #3: Unaccountability.** NSF began the now widespread practice of making grants to universities and other non-profit institutions, with scientists, usually faculty members, being classed as "principal investigators." – PIs in the argot. If PIs were directly contracted and subject to Federal Acquisition Regulations (FAR), the PIs would be directly and legally accountable. But that is not how NSF designed the system. The consequence of the long prevailing NSF methodology is that there is no direct legal responsibility or liability for the scientists' conduct. All too often scientists misrepresent with impunity the state of scientific knowledge and engage in anticompetitive practices, including science suppression, and the blacklisting of other capable, experienced scientists. University and institution administrators, when made aware of such conduct, in my experience, do nothing to correct it, having neither the expertise nor, with tenure, the perception of authority or responsibility. The result is that American taxpayers' money is wasted on a grand scale and the science produced is greatly inferior to what it might be.

•**NSF FLAW #4: For-profit Scientific Publishers.** NSF began the now widespread practice whereby the government pays the publication costs, "page charges" for scientific articles in journals run by for-profit companies or by special-interest science organizations. Because these publishers demand ownership of copyrights, taxpayers who want to obtain an electronic copy must pay, typically $40, for an article whose underlying research and publication costs were already paid with taxpayer dollars. Moreover, commercial and protectionist practices often subvert the free exchange of information, which should be part of science, making the publication of contradictions and new advances extremely difficult. Furthermore, publishers have little incentive or mechanism to insist upon truthful representations. For example, in ethical science, published contradictions should be cited, but with the extant system it is common practice to ignore contradictions that may call into question the validity of what is being published. The net result is that unethical scientists frequently deceive the general public and the scientific community, and waste taxpayer-provided money on questionable endeavors.

Ever vigilant and perhaps unrealistically optimistic, I have described these four fundamental NSF-instigated flaws that now pervade virtually all civilian U. S. Government-supported science-funding, and have proposed practical ways to correct them, which I communicated to two NSF directors, who chose to ignore them. (Appendix D) There seems to be a widespread perception of intrinsic "infallibility" in the university-government complex, wherein any action, regardless of the seriousness of its adverse consequences, is considered beyond reproach.

On December 16, 2004, an individual in the White House to whom I had complained about the inequity of "peer review" sent me a copy of the U. S. Office of Management and Budget's Final Information Quality Bulletin for Peer Review: December, 15, 2004. I sent to the White House my critique of that Bulletin and my recommendations for systemic changes, which as far as I know were neither appreciated nor implemented. (Appendix D) Ten years later, the U. S. Government still conducts "peer review" according to that Bulletin, which:

- **Embodies the tacit assumption that "peer reviewers" will always be truthful, and fails to provide any instruction, direction, or requirement either to guard against fraudulent "peer review" or to prosecute those suspected of making untruthful reviews;**
- **Approves the application of anonymity and even appears to promote some alleged virtue of its use, "*e.g.*, to encourage candor";**

- **Gives tacit approval to circumstances that allow conflicts of interest and prevents the avoidance of conflicts of interest, and;**
- **Fails to recognize or to admit the debilitating consequences of the long-term application of the practices it approves.**

As a consequence of more than half a century of badly flawed administration processes, the National Science Foundation, NASA, and other funding agencies have been doing what no foreign power or terrorist organization can do: slowly, imperceptibly undermining American scientific capability, driving America toward third-world status in science and in education, corrupting individuals and institutions, rewarding the deceitful and the institutions that they serve, stifling creative science, and infecting the whole scientific community with flawed anti-science practices based upon an unrealistic vision of human behavior. Decades of U. S. Government science-funding has produced a community of academic scientists that is willing to provide "justification" for a badly warped malevolent political agenda based upon fallacious science.

The Problem with Models

To a great extent, science ceased to be science with the widespread making of models. The American statistician, George E. P. Box, has stated this about models: "Essentially, all models are wrong, but some are useful" [1]. Models that predict the paths of hurricanes, for example, are useful and they can be improved with the data taken during each hurricane. Making a model of an event that occurred once billions of years ago or making a model in which the underlying science is unknown is neither useful nor correct.

Teachers of laboratory courses in high school and college rarely tell students beforehand the value they are trying to determine by experiment, because too often the student will manipulate the data to arrive at that known result. In making models, scientists know beforehand what they are modeling and what they think the result is supposed to be. So, the temptation is for model-makers to adjust their assumptions and the parameters to achieve the known-beforehand result; otherwise they fail in model-making. And that is what computational model-makers do.

From time to time in the 1970s, I would drive past an institute whose dedicated purpose was creation research. One day out of curiosity, I stopped by and introduced myself. I was invited to tour the facility and thus met several scientists who were engaged in efforts to attempt to show that the age of the Earth is no greater than 6,000 years, for example, by trying to show the dinosaurs were even of younger age than that. Why 6,000 years? That date arises because in 1650 Archbishop James Ussher calculated a biblical chronology which set the year of creation as 4004 B.C.

After meeting with the staff scientists, I concluded my visit to the institute by meeting with its Director. After expressing my appreciation for the courtesies extended by the research staff, I politely asked the Director to explain how he could justify their research, while ignoring the many thousands of measurements showing considerably greater age which are based upon about a dozen different radioactive dating techniques. His answer was simple and direct: "That's our model."

Geophysicists and astrophysicists like to make models based upon arbitrary assumptions, and to make models based upon other models. Even if the model appears to model what is intended to be modeled, there is no certainty that nature behaves according to the modeler's assumptions. Models do not have to be correct and usually are not. Making models, which many can do easily, is no substitute for the much more difficult task of making scientific advances by discovering fundamental quantitative relationships in nature that are securely anchored to the properties and behavior of matter and radiation. The difference between a model and what it models is like the difference between a doll and a woman. Yet models, flawed though they are, often become part of the NSF and/or NASA "accepted" storyline, which can lead to widespread confusion and misinformation. Many who make models are ill-trained in science, and believe that making models *is* science; they have little understanding of what is *not* known. Those are the individuals that can unwittingly be exploited by a malevolent political party.

Stellar Ignition

At the beginning of the 20th century, understanding the nature of the energy source that powers the Sun and other stars was one of the most important problems in physical science. Initially, gravitational potential energy release during protostellar contraction was considered, but calculations showed that the energy released would only be sufficient to power a star for a few million years [2, 3], a much shorter time than the early underestimates of the age of the Earth [4] or the time that life has existed on Earth. The discovery of radioactivity and the developments that followed, especially the discovery of nuclear fusion reactions [5], led to the idea that stars are powered by thermonuclear fusion reactions [6, 7]. (Figure 9.1)

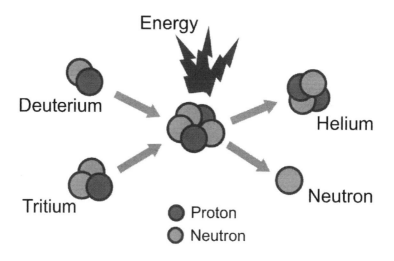

Figure 9.1 Schematic representation of nuclear fusion showing momentary the fusing-together of two light nuclides, ^{2}H and ^{3}H, which splits to yield ^{4}He, a neutron, and a burst of energy.

Thermonuclear fusion reactions are called "thermonuclear" because temperatures on the order of a million degrees Celsius are required. The principal energy released from the detonation of hydrogen bombs comes from thermonuclear fusion reactions. The high temperatures necessary to ignite H-bomb thermonuclear fusion reactions come from their A-bomb nuclear fission triggers. Each hydrogen bomb is ignited by its own small nuclear fission A-bomb.

In 1938, when the idea of thermonuclear fusion reactions as the energy source for stars was reasonably well developed [6], nuclear fission and the nuclear fission chain reaction had not yet been discovered [8]. Astrophysicists assumed that the million degree temperatures necessary for stellar thermonuclear ignition would be produced by the in-fall of dust and gas during star formation and have continued to make that assumption to the present, although clearly there have been signs of potential trouble with the concept. Proto-star heating by the in-fall of dust and gas is off-set by radiation from the surface, which is a function of the fourth power of temperature. What does this mean? In non-technical terms it is this. A star that is in the process of forming has not yet ignited to produce its own energy. But if heat is being added to it from in-falling dust and gas, it will shine at the expense of the energy that is being added. And, at each successive increase in temperature, it shines away a vastly larger amount of energy. So, even before the forming-star reaches the temperature for ignition of its thermonuclear reactions, it is already shining away a HUGE amount of energy. From the in-fall of dust and gas alone, the forming-star, I submit, may never be able to get hot enough to ignite. But that is the idea that has persisted virtually unchanged since the 1930s.

Generally, in numerical models of protostellar collapse, thermonuclear ignition temperatures, on the order of a million degrees Celsius, are not attained by the gravitational in-fall of matter without additional *ad hoc* assumptions, such as a shockwave-induced sudden flare-up [9, 10] or result-optimizing the model-parameters, such as opacity and rate of in-fall [11].

The idea of attaining stellar thermonuclear ignition temperatures from the in-fall of dust and gas was initially assumed in the early 1930s as no other feasible heat source was known; nuclear fission and the nuclear fission chain reaction had not yet been discovered. After demonstrating the feasibility for planetocentric nuclear fission reactors for the giant, gaseous planets [12], I suggested, in a paper published in the *Proceedings of the Royal Society of London* [13], and copied as Appendix E, that thermonuclear fusion reactions in stars, as in hydrogen bombs, are ignited by self-sustaining, neutron-induced, nuclear fission chain reactions. A star, after all, is like a hydrogen bomb held together by gravity.

Stellar Non-Ignition

Since the 1930s, all stars, except brown dwarfs, have been thought to ignite automatically upon formation and to be luminous. The idea that stars are ignited by nuclear fission triggers admits the possibility of stellar non-ignition in the absence of fissionable elements. This new concept may have fundamental implications bearing on the nature of dark matter [13], especially the dark matter thought to reside in galactic halos and to be responsible for dynamic stability [14]. Moreover, the idea of stellar thermonuclear fusion ignition by nuclear fission provides a new concept for internal heat production in hot Jupiter exoplanets, and for the thermonuclear ignition of dark galaxies [15]. (Appendix F)

The concept that thermonuclear fusion reactions in stars are ignited during stellar formation by temperatures produced by the heat generated from the in-fall of dust and gas originated before nuclear fission and the nuclear fission chain reaction were discovered and leads to the tacit assumption that all stars more massive than brown dwarfs automatically ignite. Most stars in the observable universe appear to be grouped into galaxies that display only a few, frequently-occurring, common morphologies (Figure 9.2), which are wholly inexplicable from the standpoint of stellar ignition by the in-fall of dust and gas. Why? Because, if stars are ignited by temperatures produced by the gravitational in-fall of dust and gas as assumed since the 1930s, then all stars, except tiny brown dwarfs, are automatically ignited during formation. Then the great question becomes how did stars arrange themselves into so few prominent galactic morphologies, such as spiral and bar galaxies? What is the mechanism? Astrophysicists cannot explain how that arrangement that might have occurred. Previously, there was no answer to that question in the scientific literature.

Observational evidence, primarily based upon velocity dispersions and rotation curves, suggests that spiral galaxies have associated with them massive, spheroidal, dark matter components, thought to reside in their galactic halos [14] which I suggested are composed of Population III stars. These zero-metallicity stars lack fissionable elements, and consequently are unable to sustain the nuclear fission chain reactions necessary for the ignition of thermonuclear fusion reactions [13]. Interestingly, a one-solar mass, Population III dark star would be about the size of Earth [16].

Figure 9.2 Hubble Space Telescope broad field image showing numerous galaxies.

The 1957 concept that elements are synthesized within stars [17] has become widely accepted by those in the government-funded astrophysics community. This concept is frequently referred to as B²FH, an acronym for the authors' initials. In the B²FH model, heavy elements are thought to be formed by rapid neutron capture, the "R-process," at the supernova end of a star's lifetime; I proffered a different explanation for the production of heavy elements, namely, the formation of highly dense nuclear matter in the galactic core and its associated jetting out into space [15].

The conditions and circumstances at galactic centers appear to harbor the necessary pressures for producing highly dense nuclear matter and the means to jet that nuclear matter out into the galaxy where the jets seed dark stars which they encounter with heavy elements, including fissionable elements, thus turning dark stars into luminous stars. Galactic jets, either single or bi-directional, have indeed been observed originating from galactic centers, although little is currently known of their nature. Figure 9.3 is a Hubble Space Telescope composite image of galactic jets, ranging in length from 4,000 to 865,000 light years.

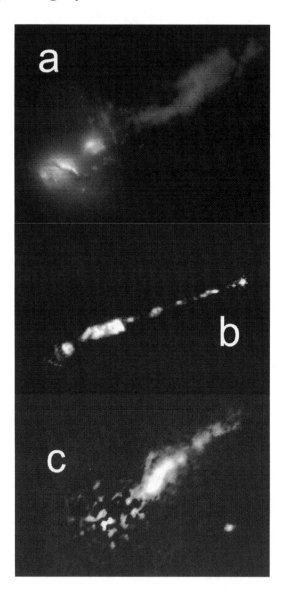

Figure 9.3 Hubble Space Telescope composite image showing three galactic jets, lengths, (a) 865,000 light years, (b) 4,000 light years, and (c) 10,000 light years.

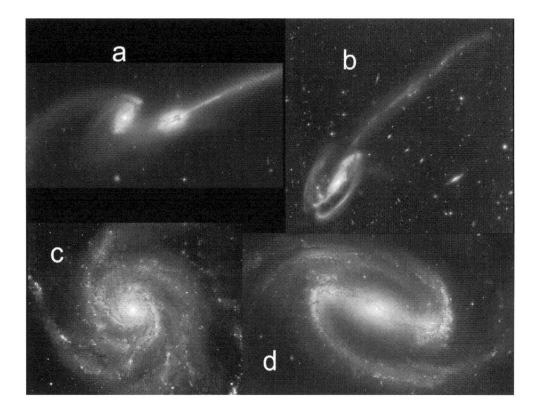

Figure 9.4 Hubble Space Telescope image of (a) anomalous galaxy NGC 4676, (b) anomalous galaxy UGC 10214, (c) spiral galaxy, M101 and (d) barred spiral galaxy NGC 1300.

Consider a more-or-less spherical, gravitationally bound assemblage of dark (Population III) stars – those consisting only of hydrogen and helium – a not-yet-ignited dark galaxy. Now, consider the galactic nucleus as it becomes massive and shoots its first jet of nuclear matter into the galaxy of dark stars, seeding and igniting those stars which it contacts. How might such a galaxy at that point appear? I suggest it would appear quite similar to NGC4676 (Figure 9.4a) or to NGC10214 (Figure 9.4b).

The arms of spiral galaxies, such as M101 (Figure 9.4c), and the bars which often occur in disc galaxies, such as in NGC1300 (Figure 9.4d), possess morphologies which I suggest occur as a consequence of galactic jetting of nuclear matter containing fissionable elements into the galaxy of dark stars, seeding the dark stars encountered with fissionable elements, thus igniting their thermonuclear fusion reactions.

The structures of just about all luminous galaxies appear to have the jet-like luminous-star features, imprints of the galactic jets which gave rise to their ignition, the imprints of the distribution of fissionable, heavy element seeds. Here is a commonality that connects a diverse range of observed galactic structures.

And what of the dark matter necessary for dynamical stability? The dark matter is the spherical halo of un-ignited, dark stars, located just where it must be to impart rotational stability to the galactic luminous structure [14].

I suggested that the distribution of luminous stars in a galaxy, and consequently the type of galaxy, for example, barred or spiral, may simply reflect the distribution of the nuclear matter containing fissionable elements jetted from the galactic center.

My concept of the thermonuclear ignition of stars by nuclear fission has not been commented upon, considered or criticized by the model-making astrophysicists. Ignoring work that challenges the "politically correct" consensus-approved storyline is common practice in the politically-driven university-government complex, thanks to the fear of retribution by secret "peer reviewers" or to the fear of being "denounced" and blacklisted.

In 2006, I submitted a short manuscript on the thermonuclear ignition of dark galaxies to *Astrophysical Journal Letters*. I signed the required copyright transfer form, and the manuscript went out for secret "peer review," but it was rejected without any substantive scientific criticism. So I submitted two other brief manuscripts. The fact that I was never asked to sign the copyright transfer forms for those other two papers prior to review as required, a serious breach of journal policy, was clear indication that my manuscripts were not going to be accorded the fair and impartial consideration that is supposed to be the usual policy of the American Astronomical Society, the journal's sponsor. Not surprisingly, those two manuscripts were rejected, this time with lengthy reviews (more words than the manuscripts) describing a litany of model-making issues, without answering the two very appropriate questions: How unique and how important are the ideas? I complained to the officers of the American Astronomical Society, who never responded, even though the bylaws of the American Astronomical Society (*AAS*) clearly state:

> **As a professional society, the AAS must provide an environment that encourages the free expression and exchange of scientific ideas.**

In rejecting those manuscripts, the American Astronomical Society hid from its members, from the scientific community, and from U. S. Government science funding officials, new, fundamental, and potentially important insights about the universe, including the basis of why galaxies have the characteristic appearances they are observed to have. Appendix F reprints the scientific article I published in *Current Science*, entitled "New Concept for Internal Heat Production in Hot Jupiter Exo-planets, Thermonuclear Ignition of Dark Galaxies, and the Basis for Galactic Luminous Star Distributions" [15].

Not long after the *Astrophysical Journal Letters* incident, on July 27, 2007, when attempting to post a preprint in the astrophysics category, I found myself blacklisted by *arXiv.org*, an author self-posting archive that functions, in theory at least, to preserve and make available to other scientists valuable potential insights that would otherwise be quashed by peer review and other bureaucratic restrictions and machinations.

The website *archivefreedom.org* displays case histories of some of the individuals who have been blacklisted by the *arXiv.org* administration and its secret moderators, and includes a statement by blacklisted scientist Nobel laureate Brian D. Josephson explaining the meaning of blacklisting as applied to *arXiv.org*. Being blacklisted by *arXiv.org* means that either your attempts to post scientific papers are disallowed, or they are "buried," which is to say re-directed to (or only permitted in) categories where scientists or mathematicians in the specific area will likely not see them, such as in General Physics or in General Mathematics. I adopt that definition of blacklisting in this book.

Before this blacklisting, I was not only permitted to automatically post papers, but also to endorse others to be able to post papers in the archive in the following categories: Astrophysics, Educational Physics, General Physics, Geophysics, History of Physics, and Space Physics. Now, for no legitimate reason, I am blacklisted, stripped of the ability to endorse others, and suffer having my scientific papers "buried" in General Physics where it is unlikely they will be noticed; that is, if they are allowed to post at all. Even my scientific papers that call into question U. S. Government funded investigations at Cornell University, the present owner of arXiv.org, are either "buried" or forbidden to post in this author-self-posting archive, *where hundreds of thousands of papers post automatically without human intervention.*

One consequence of NSF's invention of anonymous "peer review" is that publication of scientific papers is often delayed for years or prevented by so--called "peer reviews" from competitors, whose primary aim is to debilitate or eliminate their competition. In the 1990s, the National Science Foundation funded the development at Los Alamos National Laboratory of an author-self-posting archive, where physicists and mathematicians could post their preprints, without interference from their competitors, making them available worldwide almost instantly. That archive underwent various name changes, eventually becoming *arXiv.org*.

Since its inception, *arXiv.org* has become the preeminent means of scientific communication in the areas of science and mathematics it hosts. Rather than wade through the many hundreds of individual scientific journals, often having limited access without paying fees, scientists can receive by email a list of daily postings in specific areas of the scientific disciplines hosted by *arXiv.org* and can download scientific articles of interest without charge. The development of the author self-posting archive might have become the jewel in NSF's crown, one of its greatest achievements. Instead, rather than conveying the archive to a neutral entity, such as the Library of Congress, NSF's mal-administration permitted it to become an instrument for science suppression and for blacklisting and discrimination against competent, well-trained scientists worldwide.

On or about 2001, key personnel responsible for developing the author-self-posting archive at Los Alamos National Laboratory left that organization to become employed by Cornell University. Presumably in a coordinated way, Cornell University, through a proposal to the National Science Foundation (NSF #0132355, July 16, 2001), took over ownership of the author-self-posting archive, now called *arXiv.org*, and presumably was given the requested $958,798 to do that. That proposal contains the following statement made to justify Cornell University's proposed use of a "refereeing mechanism":

> **The research archives become less useful once they are inundated for example by submissions from vociferous "amateurs" promoting their own perpetual motion machines....**

One consequence of *arXiv.org* blacklisting is to deceive U. S. Government science funding officials and individuals conducting scientific investigations and teaching science, keeping them in the dark about new ideas and discoveries. Beyond the financial and professional debilitation suffered by blacklisted scientists and mathematicians, there is also a human toll. One blacklisted individual, Florentin Smarandache, notes [18]:

Blacklisted scientists are subject to derision, ignorance, insults, lies, false accusations, personal attacks against them, misrepresentations regarding their research, culture, faith, etc.

Hundreds of thousands of scientific papers have been posted on *arXiv.org* without any human intervention at all. Human intervention, but *not* "peer review," occurs *only* when an individual is "denounced," intentionally singled out for disparate treatment, through the application of unfair, arbitrary, and capricious standards. Being tagged for disparate human intervention, just as being denounced in a totalitarian state, may occur for any number of unspecified reasons. Human intervention is perpetrated by *arXiv.or*g administrators in conspiracy with a small group of *arXiv.or*g "insiders" who may or may not call themselves "moderators" and who discriminate in secret and without any accountability. Moreover, there is no recourse: In my experience, Cornell University's librarian, provost and president do nothing to provide oversight responsibility for the conduct of *arXiv.org*, typically referring complaints back to the *arXiv.org* administrators who are the subject of the complaint in the first place. Being "denounced" for disparate treatment by secret "insiders," without recourse, is something I might have expected from the now-defunct Soviet Union or from Ceausescu's Romania. But, here it is in America; bought and paid for by the National Science Foundation. As an American citizen, veteran, and taxpayer, I am justifiably appalled by this perverse, Kafkaesque misuse of public funds.

In my view, there is something fundamentally wrong with Cornell University receiving U. S. Government grants and contracts to conduct scientific research, and then deceiving the scientific community, via *arXiv.org*, by not posting or by hiding new advances or contradictions, especially in instances that potentially impact the investigations being performed at government expense at Cornell. Cornell University is a recipient of millions of dollars in U. S. Government grants and contracts, and is one of a pool of competitors for Federal grants and contracts. The U. S. National Science Foundation, I submit, made a major blunder, with serious anti-scientific results, in turning over to Cornell University a powerful tool (*arXiv.org*) that could be used against its competitors. In doing so, I allege, the National Science Foundation violated the very law that created NSF:

In exercising the authority and discharging the functions referred to in the foregoing subsections, it shall be an objective of the Foundation to strengthen research and education in the sciences and engineering, including independent research by individuals, throughout the United States, and *to avoid undue concentration of such research and education. 42 United States Code 1862 (e)* [italics added]

Instead of obeying that law, the U.S. National Science Foundation placed into the hands of one major, well-financed competitor a powerful tool (*arXiv.org*) that could not only be applied arbitrarily with capricious standards against its competitors, but by such actions casts a shadow of fear at being "denounced" in secret and there upon being blacklisted, further ensuring conformity and science suppression.

Carlos Castro Perelman, a U. S. citizen, earned his B.S. degree in physics from M.I.T. and Ph.D. from the University of Texas. Perelman has published numerous scientific papers in recognized peer-reviewed journals. He states [19]:

when I tried submitting my most recent paper (at the time) in early February 2000 to the hep-th (high energy physics theory) category my paper was removed and displaced to the general physics category (the bottom of the pile in readership and audience).

That was the beginning of Perelman's archive blacklisting; he became fully blacklisted in early 2003, in other words, not allowed to post preprints at all. Whoa! What's wrong with this picture? The date of Perelman's first blacklisting was 10 months *before* Cornell University's NSF proposal #0132355 with the "vociferous amateurs" justification for proposed use of a "refereeing mechanism." Furthermore, documents posted on *archivefreedom.org* indicate that the Director of Los Alamos Scientific Laboratory as well as NSF personnel were aware of another individual, Robert Gentry, being blacklisted at least four months before the date of submission of that proposal with the "vociferous amateurs" justification. Cornell University, I allege, misled the National Science Foundation with said "vociferous amateurs" justification as evidence appears to indicate that, prior to the submission date of said proposal, highly educated scientists were being subjected to arbitrary and capricious treatment by archieve personnel.

To date more than 988,000 scientific papers have been self-posted on *arXiv.org* since its inception in the 1990s. Generally, anyone whose email address ends in .gov or .edu automatically is allowed to post; others must first obtain the endorsement from someone who has met the requirements to endorse: That endorsement system in itself should preclude "*submissions from vociferous "amateurs" promoting their own perpetual motion machines.*" No one associated with that archive has either the time or expertise to "referee" or "adjudicate" the correctness or appropriateness of those papers. Refereeing was never the intended purpose of the archive. The original intent of the archive was to allow scientists to post pre-prints of their papers *without interference*, even before submitting them to journals, so that the work would almost instantly become available to other persons working in their particular areas of science. Evidence indicates, however, that said original intent was selectively being perverted from the beginning of the operation of the archive through to the present by the application of unfair, arbitrary, and capricious standards. And to the claim of "vociferous," I plead guilty; although a glance at Appendix G testifies that I am not an amateur.

Cornell University, its Provost, its University Librarian, and those who administrate *arXiv.org*, I allege, engage in practices that unreasonably restrain competition in violation of antitrust laws and engage in exclusionary acts that maintain monopoly power in violation of antitrust laws and they do so and/or have done so, in part using resources and/or financial support provided by the U. S. Government.

Most of the support for basic scientific research in the United States comes from grants and contracts provided by the U. S. Government. With very rare exceptions, the grant-funding mechanisms are essentially identical in the various funding agencies. Proposals are submitted to the relevant agency, such as the National Science Foundation, in discipline-specific programs and are subjected to agency review and to peer review in the same discipline-specific areas. For example, a proposal to study the formation of galaxies would be submitted to the astrophysics program, not to the geophysics program, and would be peer-reviewed by scientists in the astrophysics community, not by those in the geophysics community.

Posting scientific papers on *arXiv.org* is advertising. Similarly, publishing scientific papers in journals is advertising, although a much less effective form of advertising. In fact, paid-for papers (most are usually paid for) in one journal, the *Proceedings of the National Academy of Sciences* carries the following printed notice:

> The publication costs of this article were defrayed in part by page charge payment. This article must therefore be hereby marked "*advertisement*" in accordance with 18 U.S.C. §1734 solely to indicate this fact.

Without access to such advertising, a scientist is essentially "non-existent" in a particular discipline-specific community and at an extreme disadvantage for being recognized by funding-agency officials or by peer-reviewers in that particular discipline. The effectiveness of *arXiv.org* as an advertising medium is described well by a quote from said proposal [NSF #0132355]:

> In a feature article Physicists in the New Era of Electronic Publishing in the August 2000 Physics Today, current American Physical Society president James Langer begins: "some of my colleagues in Santa Barbara — the string theorists, for example, and several of my coworkers in condensed matter theory as well — insist that they don't need The Physical Review. For research purposes, they don't need refereed print journals at all. They are producing remarkable results this way, so I take them very seriously. What they are doing is using the Los Alamos e-print archive for all of their research communications. They check it every day for new information. They post all their papers there, cite references by archive number, use the search engine to find other papers, and need little or no other publication services. Publication on the archive is instantaneous. It costs the users nothing and is self-organizing — or at least it appears so. It's also far more democratic than the old system with which I grew up. Physicists all over the world can post their research results without being hassled by grumpy editors and

referees. And they don't have to be part of some inner circle of accepted colleagues to be on the preprint mailing lists; they can find out what's new on the archive just as soon as everyone else does...."

Consider the following two hypothetical cases: Prime-time television advertising is the life-blood of American automobile manufacturers. Suppose a new automobile manufacturer meets all the requirements and desires to advertise on prime-time television, but a secret group of conspirators with vested interests prevents the manufacturer either from advertising on television [Case 1] or from advertising at any time other than in the early morning hours on children's programs [Case 2]. Clearly, such practices would unreasonably restrain competition in violation of antitrust laws. An identical parallel has been happening on *arXiv.org* since its inception and continues to the present, but with one fundamental difference: Throughout most of its existence, *arXiv.org* has been supported fully or partially by the U. S. Government through grants from the National Science Foundation and/or by grants from **other tax exempt organizations, notably, the Simons Foundation and/or its member institutions.**

The *arXiv.org* parallel to that hypothetical example is the following: [Case 1] a secret group of *arXiv.org* insiders with vested interests unreasonably prevent a targeted individual from posting on *arXiv.org* by not allowing posting, or simply by making attempted postings disappear; [Case 2] a secret group of *arXiv.org* insiders with vested interests unreasonably prevent a targeted individual from posting in the desired and appropriate discipline-specific area on *arXiv.org*, wrongfully and unwarrantedly re-categorizing the paper to some inappropriate category, *i.e.*, "burying" it. The actions in Case 1 and in Case 2 are or have been accomplished using U. S. Government provided resources and/or financial support and/or with support from tax exempt organizations.

There is an *arXiv.org* process for becoming qualified to endorse others, and for being stripped of the ability to endorse, if one makes fallacious endorsements. I never endorsed anyone, but by arbitrary and capricious standards, I was stripped of the ability to endorse. Being unwarrantedly stripped of the ability to endorse others adversely distorts *arXiv.org* advertisements, visible representations of professional capability and expertise, thereby diminishing the advertising impact. (Figure 9.5)

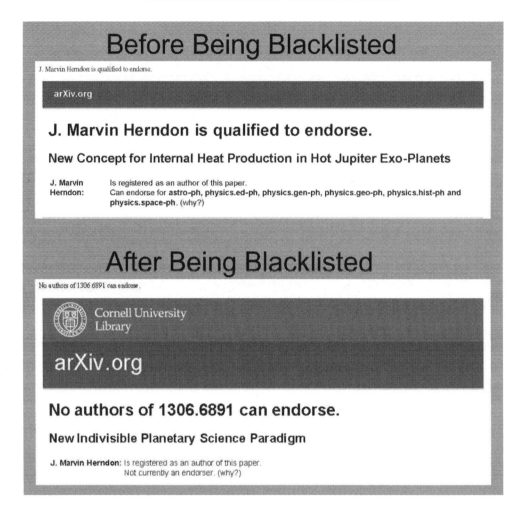

Figure 9.5 Comparison of publicly available *arXiv.org* advertising before and after blacklisting.

Disparate treatment and the application of unfair, arbitrary, and capricious standards by the administrators of *arXiv.org*, even when their purposes are to unreasonably restrain competition and/or to achieve or maintain monopoly power, I allege, violate the civil rights of individuals as well. Among the plethora of *arXiv.org* blacklisted individuals, there may be some who are being discriminated against for other than the anti-trust related reasons described above, namely, maintaining a monopoly and/or restraining competition.

It might be difficult or impossible, for example, to determine whether the *arXiv.org* blacklisting of Carlos Castro Perelman, Ph.D. [19] is related in some way to his being Hispanic or to his former affiliation with Clark Atlanta University, a historically black institution of higher education. Perelman [19] quotes a statement in an email communication by Paul H. Ginsparg, Principle Investigator of said proposal. In that email Ginsparg uses the following phrases:

> **castro [sic] is an obvious nut and all of his papers are abject nonsense... castro [sic] is more than welcome to publish in conventional journals we don't have time for him here, and he is fortunate that he is permitted in the General Physics category.**

The phraseology and tone of those remarks might seem to indicate a personal and deep seated hatred of Carlos Castro Perelman or, perhaps, what he appears to represent. Other examples are posted on *archivefreedom.org* [18].

Often psychological pain accompanies blacklisting, as well as the risk of being cheated of the proper recognition and economic advantage for being the original thinker of an idea that may gain appreciation in the future. Being blacklisted may result in a scientist's reputation becoming tarnished, and could lead to loss of potential grants, job opportunities, and resulting income. All of this occurs because of the intentional application of unfair, arbitrary, and capricious standards by the administrators of *arXiv.org*, allowing a small, largely secret, clique to practice discrimination with ease, with impunity, without accountability, and without recourse, while in part using or having used resources and/or financial support provided by the U. S. Government and/or by tax exempt organizations.

I have filed formal complaints with voluminous documentation to Cornell University Provosts **Carolyn A. Martin** and Kent Fuchs. That I am still blacklisted by *arXiv.org* and that *arXiv.org* operates essentially unchanged demonstrates, I submit, approval of current blacklisting practices by those Officers of Cornell University.

On June 30, 2008 I wrote and provided evidence to Democrat Andrew M. Cuomo, then Attorney General for the State of New York. My letter began:

> **I herewith request a formal investigation into allegations of black-listing, unlawful restraint of trade, and violations of individuals' civil liberties by the management of the author-self-posting archive, arXiv.org, the University Librarian, and the Provost of Cornell University.**

I received no response.

On July 23, 2008 I wrote to Democrat U. S. Senator Barbara Boxer of California making a similar request. Again I received no response.

After a considerable passage of time with no response from Boxer, I forwarded a copy of that letter to Republican Congressman Duncan Hunter of San Diego, California. In the early months of Democrat Barack Obama's presidency, Republican Congressman Hunter forwarded allegations of wrong-doing and dozens of evidentiary documents to Assistant Attorney General for Anti-trust Garza. The U. S. Department of Justice did nothing. Unfortunately, I am reminded of the following words [20]:

> **Where we are there is no place for anyone else! –**
> *Adolph Hitler*

Cornell University is a microcosm of American universities. A research report, entitled "Political Bias in the Administrations and Faculties of 32 Elite Colleges and Universities," found that the overall ratio of identifiable Democrats to Republicans at the 32 schools was more than 10 to 1 (1397 Democrats, 134 Republicans) [21]. The report noted that administrators lean just as far to the left; at Cornell University not a single Republican administrator could be identified. Another investigation reports similar bias and notes, "Faculty at colleges and universities of all kinds in America are overwhelmingly liberal in their political ideology, creating a strong campus political culture" [22]. If faculty at universities and colleges were as overwhelmingly Republican instead of Democrat for half a century, then I suspect there might be a different, but parallel horror story. When has such ideological dominance ever had a happy ending? But here the figures clearly speak to the Democrat politicization and distorted scientific group think of the modern university. Overwhelming Democrat dominance of universities has weakened geoscience integrity to the point of willingly and unquestioningly serving the malevolent political agenda described in Chapter 1.

Chapter 9 References

1. Box, G.E.P., *Empirical Model-Building and Response Surfaces*1987: Wiley.

2. Helmholtz, H.v., *On the interaction of natural forces*. Phil. Mag., 1854. 11: p. 489-518.

3. Thompson, W., *On the mechanical energies of the Solar System*. Phil. Mag., 1854. 8: p. 409-430.

4. Joly, J., *An estimate of the geological age of the Earth*. Trans. R. Soc. Dublin, 1899. VII: p. 23-66.

5. Oliphant, M.L., P. Harteck, and E. Rutherford, *Transmutation effects observed with heavy hydrogen*. Nature, 1934. 133: p. 413.

6. Bethe, H.A., *Energy production in stars*. Phys. Rev., 1939. 55(5): p. 434-456.

7. Gamow, G. and E. Teller, *The rate of selective thermonuclear reactions*. Phys. Rev., 1938. 53: p. 608-609.

8. Hahn, O. and F. Strassmann, *Uber den Nachweis und das Verhalten der bei der Bestrahlung des Urans mittels Neutronen entstehenden Erdalkalimetalle*. Die Naturwissenschaften, 1939. 27: p. 11-15.

9. Hayashi, C. and T. Nakano, *Thermal and dynamic properties of a protostar and its contraction to the stage of quasi-static equilibrium*. Prog. theor. Physics, 1965. 35: p. 754-775.

10. Larson, R.B., *Gravitational torques and star formation*. Mon. Not. R. astr. Soc., 1984. 206: p. 197-207.

11. Stahler, S.W., et al., *The early evolution of protostellar disks*. Astrophys. J., 1994. 431: p. 341-358.

12. Herndon, J.M., *Nuclear fission reactors as energy sources for the giant outer planets*. Naturwissenschaften, 1992. 79: p. 7-14.

13. Herndon, J.M., *Planetary and protostellar nuclear fission: Implications for planetary change, stellar ignition and dark matter*. Proceedings of the Royal Society of London, 1994. A455: p. 453-461.

14. Rubin, V.C., *The rotation of spiral galaxies.* Science, 1983. 220: p. 1339-1344.

15. Herndon, J.M., *New concept for internal heat production in hot Jupiter exo-planets, thermonuclear ignition of dark galaxies, and the basis for galactic luminous star distributions.* Current Science, 2009. 96: p. 1453-1456.

16. Lynden-Bell, D. and J.P. O'Dwyer, *One mass-relation for planets, white dwarfs, and neutron stars.* arXiv.org/abs/astroph/0104450, 2008.

17. Burbidge, E.M., et al., *Synthesis of the elements in stars.* Rev. Mod. Phys., 1957. 29(4): p. 547-650.

18. ArchiveFreedom.org.

19. Perelman, C.C., *My Struggle with Ginsparg (arXiv.org) and the road to cyberia: A scientific-gulag in cyberspace,* in *Against the Tide: A Critical Review by Scientists of How Physics & Astronomy Get Done,* C.C.P. Martin Lopez Corredoira, Editor 2008, Universal Publishers: Baca Raton, Florida. p. 59-76.

20. Delarue, J., *The Gestaop: A History of Horror*1962, New York: Skyhorse Publishing.

21. Horowitz, D. and E. Lehrer, *Political Bias in the Administrations and Faculties of 32 Elite Colleges and Universities,* 2003, Students for Academic Freedom.

22. Tobin, G.A. and A.K. Weinberg, *A Profile of American College Faculty: Volume I: Political Beliefs and Behavior*2006: Institute for Jewish & Community Research.

10. A Matter of Responsibility

The whole art of government consists in the art of being honest. — *Thomas Jefferson*

Truth never damages a cause that is just. — *Mahatma Gandhi*

Truth is a pillar of civilization. The word "truth" occurs 224 times in the King James Version of the *Holy Bible*. Witnesses testifying in American courts and before the United States Congress must swear to tell the truth. Laws and civil codes require truth in advertising and in business practices. Truthfulness is important in virtually every activity and circumstance, not just in jurisprudence. Truth and trust are inextricable in human experience, in families and in relationships. And in science, perhaps more than any other activity, truth and the search for it is paramount. In fact, good science is inconceivable without an extremely high value being put on truth. Responsibility for scientists begins with telling the truth, the full truth, a truth that transcends personal animus, cultural differences, corpo-political pressures; a truth that permeates all other responsibilities. But, if you have read this far in this book you will be familar with my analysis of how the geosciences have become disconnected both from the necessity of being truthful and from a sense of responsibility to society as a whole, at least in America and now increasingly in Western Europe.

On November 17, 1944, U. S. President Franklin D. Roosevelt (1882-1945) wrote to Vannevar Bush (1890-1974), Director of the Office of Scientific Research and Development, asking for recommendations concerning post-World War II support for science, stating in part:

> **The information, the techniques, and the research experience developed by the Office of Scientific Research and Development and by the thousands of scientists in the universities and in private industry, should be used in the days of peace ahead for the improvement of the national health, the creation of new enterprises bringing new jobs, and the betterment of the national standard of living…. New frontiers of the mind are before us, and if they are pioneered with the**

same vision, boldness, and drive with which we have waged this war we can create a fuller and more fruitful employment and a fuller and more fruitful life.

On July 25, 1945, Vannevar Bush responded to U. S. President Harry S. Truman (1884-1972), transmitting the report *Science, the Endless Frontier*, which was to become the blueprint for peacetime U. S. Government support for science. That document provided the inspiration and justification for the 1951 establishment of the U. S. National Science Foundation (NSF) by the U. S. Congress.

Scientists by virture of their ability and training possess, or are often perceived to possess, special insight into the nature of Earth and universe. During the first half of the 20th century, scientists indeed achieved great prestige and earned high respect for their fundamental discoveries and for their strict adherence to the truth. But that apparently, and unfortunately, changed with the establishment of the National Science Foundation. As I described in Chapter 9, the NSF institutionalized a deeply flawed science-administration process that has, however unintentionally, trivialized and corrupted scientific research while rewarding the deceitful, those who deceive scientists and the public and who all too often, to promote their own interests from behind a protective veil of secrecy, lie as "peer reviewers." And NSF has compounded the problem by bringing in individuals from grant-recipient institutions to serve for periods as assistant program managers. That process, like NASA's Lunar and Planetary Geoscience Review Panel (Chapter 8), helped to forge an unholy alliance between the taker and the giver of taxpayer provided grant-monies that I refer to as the university-government complex.

In his 1961 Farewell Address, U. S. President Dwight D. Eisenhower (1890-1969), famous for warning of the dangers of the military-industrial complex, warned as well of the adverse influence of Federal funding on science:

> **Today, the solitary inventor, tinkering in his shop, has been overshadowed by task forces of scientists in laboratories and testing fields. In the same fashion, the free university, historically the fountainhead of free ideas and scientific discovery, has experienced a revolution in the conduct of research. Partly because of the huge costs involved, a government contract**

becomes virtually a substitute for intellectual curiosity…. The prospect of domination of the nation's scholars by Federal employment, project allocations, and the power of money is ever present and is gravely to be regarded.

Looking into the future is always difficult, but President Eisenhower's prognosticatory powers were not lacking. He had foreseen clearly a major institutional problem. But his greatest concerns never envisioned the consequential endgame, the structural group-think of scientists in the university-government complex who are now so cowed that it might be career or reputation suicide to even consider alternatives to, for example, the anthropogenic global warming malevolent political agenda that now appears to pervade virtually every aspect of American governance (Chapter 1). As I write this chapter, decapitating barbarians, some funded directly and indirectly by the West, are on the move in the Middle East and threaten America; Ebola may pose a potential pandemic; meanwhile, the U. S. Department of Defense (DOD) focuses on adapting to "anthropogenic climate change" as described in an October 15, 2014 DOD Report "2014 *Climate Change Adaptation Roadmap.*" Why? Money? Orders from above?

President Eisenhower was correct about the "power of money," but more generally I would add the *taking of other peoples' money* because that is one of the two major driving-forces behind individuals and institutions joining the global warming malevolent agenda; the other is *control over people and industries* by none other than those whose very existence depends upon *taking of other peoples' money.* These driving-forces are common to virtually all malevolent political agendas, including those of Germany and Japan in World War II.

Forces of Deception

The global warming agenda had its beginnings in the 1990s, but from my experience institutionalized deceit in the geoscience community goes back considerably further. Paul K. Kuroda was concerned about anonymous reviewers in 1956; my experience began in 1979 with the systematic ignoring of the new idea [1] about the composition of the Earth's inner core and with my subsequent "excommunication." (Appendix A) Even at the time of its publication in the *Proceedings of the Royal Society of London*, there were obvious potentially game-changing implications for geophysics. The early inner-core idea [2], proposed circa 1940 and still extant, implies that the core temperature is constant: any heat added or removed would cause the inner core to shrink or to grow, and the temperature would remain constant until either the core was fully solidified or the inner core had fully melted. Under WEDD, however, if the inner core is fully crystallized nickel silicide, as I suggest, the addition or removal of heat would change the core temperature. The entire thermal structure of the Earth is in question. A real scientist would want to know which, if either, is correct. A real scientist would also want to explore the possibility, the fundamental implication of a radically different Earth origin with the associated suggestion that major amounts of uranium may exist in its core [3]. So why was my original and non-disgruntled work singled out, as I have chronicled, for systematic exclusion?

Already, thirty five years ago, thanks to NSF policies, there had developed a geoscience "Mafia" (no connection to La Cosa Nostra); the forces that have so successfully perverted university and government science have been honed and perfected for a long time, and are truly pervasive.

Science societies, like trade organizations, are generally composed of member scientists of various specialties. Typically, each of the various science specialties and sometimes even sub-specialties has an organization to represent its members. The American Astronomical Society, mentioned in Chapter 9, is one example. In the U. S., the American Geophysical Union (AGU) and the American Physical Society (APS) have long been proponents of anthropogenic global warming [4].

Typically, professional societies have Codes of Conduct that define clearly the ethical conduct and practices that scientists of integrity, including the organization itself, are expected to follow. These are echoed in the following excerpts from The Chemical Professional's Code of Conduct of the American Chemical Society:

> **Chemical professionals should seek to advance chemical science, understand the limitations of their knowledge, and respect the truth. They should ensure that their scientific contributions, and those of their collaborators, are thorough, accurate, and unbiased in design, implementation, and presentation . . . maintain integrity in all conduct and publications, and give due credit to the contributions of others. Conflicts of**

interest and scientific misconduct, such as fabrication, falsification, and plagiarism, are incompatible with this Code.

I have been a member of the American Geophysical Union since 1973. The May 2, 1995 issue of *Eos, Transactions, American Geophysical Union*, contained an article entitled "Neptune's Nemesis" that described observations of a new dark spot in the atmosphere of the planet Neptune. In addition to having a historical error, the article failed to represent to the geophysics community the significance of the observation with respect to possible on-going changes in the internal energy source, presumably, a central nuclear fission reactor that drives Neptune's turbulent atmosphere. I responded with a brief manuscript, and specifically requested no editorial involvement by personnel from NASA's Jet Propulsion Laboratory (JPL) or to the California Institute of Technology (Caltech) that operates JPL for NASA because of a possible institutional conflict of interest. In blatant contradiction to AGU's written policy on the avoidance of real or apparent conflicts of interest, I was told that an employee of JPL would serve as editor and that my only other option was to withdraw the article. Not surprisingly, the JPL-based editor demanded as a condition for publication that I remove all mention of the significance of the observations with respect to possible on-going changes in Neptune's internal energy source which included references to my published work on planetary nuclear fission reactors [5-7]. I appealed to the *Eos* Editor-in-Chief, AGU's Executive Director, Fred Spilhaus, who, without ever providing substantive scientific criticism by competent referees, reiterated the JPL-based editor's demand and falsely asserted that my manuscript contained unpublished *"original hypotheses regarding nuclear energy source"* and *that "Eos is a news and review journal, not a publication avenue for new scientific and technical ideas, which should be submitted for publication to a scientific journal."* Even when advised that the manuscript contains no unpublished "original [by which they meant unpublished] hypotheses," but was based upon published work [5-7], the Editor-in-Chief nevertheless persisted in his demand for censorship. Next, I appealed to the AGU president, who at the time was also an employee of NASA's JPL. Not surprisingly, she denied my appeal and "closed the case."

AGU's Executive Director, A.F. (Fred) Spilhaus, Jr., served in that position for 39 years during which time he undoubtedly exerted considerable influence over that organization and over the community of geophysicists. In

2010 he was replaced by a woman, Christine W. McEntee from outside the geoscience community. One of her early actions was to set up a blue-ribbon "Task Force on Scientific Ethics and Integrity"; AGU members were invited to submit comments.

On February 2, 2012, I sent the Task Force a lengthy document defining scientific ethics/integrity, which included the above American Chemical Society excerpts. I also included numerous descriptions of breaches of scientific ethics/integrity by AGU "luminaries." Just 18 days later, in the face of national exposure, the Task Force Chairman, a prominent global warming activist, Peter Gleick, admitted in disgrace that under a different name he had secured sensitive documents from the Heartland Institute, an organization opposed to global warming, and had distributed those documents to others by email, presumably to embarrass the Heartland Institute. Chairman Gleick resigned; so much for ethics/integrity at AGU.

The American Geophysical Union and the American Physical Society, like many other professional organizations, publish scientific journals and benefit financially from page charges and subscriptions that are often paid from DOE, NASA, or NSF grants. Through selection of editors, those organizations exert considerable control over the scientific literature. *Physical Review Letters*, an APS journal, rejected the paper of renowned Bell Laboratories neutrino physicist, R. S. Raghavan, entitled "Detecting a Nuclear Fission Reactor at the Center of the Earth" [8]. The same journal rejected without a substantive scientific basis my manuscript on the thermonuclear ignition of stars, which was subsequently published in the *Proceedings of the Royal Society of London. Physical Review Letters* likewise rejected my manuscript showing the physical impossibility of mantle convection. My appeals got nowhere.

On September 13, 2011, Nobel laureate Ivar Giaever, an APS Fellow, resigned in protest from the American Physical Society objecting to the following APS statement:

> **The evidence is incontrovertible: Global warming is occurring. If no mitigating actions are taken, significant disruptions in the Earth's physical and ecological systems, social systems, security and human health are likely to occur. We must reduce emissions of greenhouse gases beginning now.**

Both the American Geophysical Union and the American Physical Society have dogmatically signed on to global-warming agendas, demonstrating a lack scientific curiosity and objectivity; and both organizations exert considerable influence thanks to their cash-cow publications. It appears to me that the AGU would like to provide guidance to the U. S. Government on climate change, but what legitimate basis does it or its members have to do that? AGU and its members have systematically misled the scientific community for decades.

New Geoscience Frontier

While geoscientists in America and in Western Europe have been making models in support of the malevolent political agenda initiated by the United Nations' Intergovernmental Panel on Climate Change (IPCC), what should they have been doing instead? I say that geoscientists should make discoveries to understand the true nature of Earth — its origin, composition and behavior — so as to be able to warn the population of impending dangers. That is the course I chose, and that is what this book is all about.

The discoveries I made over a period of thirty five years are integrally related and form the basis of a new indivisible geoscience paradigm called Whole-Earth Decompression Dynamics (WEDD) that includes the following highlights:

- Earth formed from a giant gaseous protoplanet coeval with condensing at high-pressures and high-temperatures with the liquid iron core forming first, followed by the rocky matter.
- Fully condensed Earth formed a Jupiter-like gas giant, its highly-reduced rocky kernel compressed to about 66% of present diameter by ~300 Earth-masses of primordial gases and ices.
- As the Sun ignited, T-Tauri phase eruptions stripped away Earth's gases and ices.
- Presumably, both before and after the T-Tauri solar eruptions, oxidized matter from the outer reaches of the Solar System added to the Earth's outer layers.
- Earth below the seismic boundary at 660 km depth resembles an enstatite chondrite, not an ordinary

chondrite, with seismic discontinuities being compositional, not the result of changes in crystal structure.

- Over time pressures built within the Earth, driven by the stored energy of protoplanetary compression and by the energy of nuclear fission and radioactive decay.
- Eventually pressures became sufficient to begin to crack Earth's contiguous, rigid shell of continental-rock.
- Decompression-driven increases in planetary volume necessitate: 1) additional ocean-floor surface area created by crack formation and crack infilling, and 2) surface curvature adjustments by the formation of fold-mountains and by the initiation of fjords and submarine canyons.
- Earth's magnetic field is generated by a central nuclear fission georeactor that also produces heat channeled to surface hotspots.
- Stored energy of protoplanetary compression, in addition to driving planetary decompression, also emplaces heat at the base of the crust.
- Currently increasing rates of earthquake numbers and/or intensities are understandable as a consequence of variations in the two dominant Earth energy sources.
- Potentially variable seawater temperature and concomitant variability in atmospheric carbon dioxide are likewise understandable as a consequence of variations in the two dominant Earth energy sources.

The fundamental differences of this new geoscience paradigm necessitate re-evaluation of virtually all aspects of the geosciences, including and especially, the underpinnings of anthropogenic global warming, and the consequential actions of the malevolent political agenda with its inherent dangers to Earth's species.

Impending Dangers

The Earth's magnetic field, as described in Chapter 6, probably arises from an Elsasser-type dynamo operating within the radioactive waste sub-shell of the georeactor. That location provides regulated thermal convection for sustained operation. The georeactor, however, is only one ten-millionth the mass of Earth's fluid core. Such a small mass admits the possibility of relatively rapid geomagnetic reversals – timescale of two months to one hundred years inferred from paleomagnetic data [9-11]. In principle, geomagnetic reversals may be triggered by great Earth trauma or by super-intense solar outbursts inducing convection-interrupting heat-producing electric current into the georeactor. Humanity is wholly unprepared for such an event although the indications, weakening geomagnetic field and increased polar wander, point to its likelihood at some time in the future.

The nuclear fission georeactor consumes uranium and can run out of fuel. When it does, the geomagnetic field will collapse, but then would never re-establish itself as it has done during its reversals. As described in Chapter 6 and in the scientific literature [12], there is helium evidence of an impending georeactor demise, but the expected time frame, under our present understanding, ranges anywhere from 100 years to 1 billion years; the uncertainty is great. The geomagnetic field is our planet's shield from the onslaught of the raging solar wind; its demise is an impending danger.

An article [13] in the October 2014 issue of *Physics Today*, a publication of the American Physical Society, begins with the statement:

> **People are causing Earth's climate to change. Natural factors like solar variability and volcanoes may have also exerted a slight warming or cooling influence recently, but they are on top of the human contribution and small by comparison.**

That statement, like the above quoted APS pronouncement that inspired Nobel laureate Ivar Giaever's resignation, is naïve, simplistic, and consequently propaganda. The nature of science is to be inquisitive; present understanding in virtually every area is subject to question, challenge, new ideas, new data, and new interpretations – that is how scientific advances are made. Strangely, even the author acknowledges that "climate change is a complex and contentious public issue" — so why the above declaration?

If we need to worry, based on evidence, there are other things to worry about. Humanity is faced with potentially a far more serious and urgent danger than from "climate change." As discussed in Chapter 1, anthropogenic climate change has become a malevolent political agenda that creates an ideal *raison d'être* to expand control over people and industries. It provides:

- **Justification for the malevolent political party, which in America, at least, has identified itself by actions and statements, as the Democratic Party, to spray potentially toxic chemicals in the air over cities and elsewhere, chemicals whose effects on people and the environment are unknown and untested;**
- **Justification for claims against America and other nations and against industries for allegedly causing damage to the environment by fossil fuel consumption;**
- **Justification for increased regulations over people and industries;**
- **Justification for increased regulations on subjects taught in schools, and;**
- **Justification to transfer money from taxpayers to the political insiders, for example, brokers of carbon credits, fiscally unsound "green energy" companies, such as bankrupt Solyndra that received $535 million in Federal loan guarantees from the Obama administration, and more.**

Unfortunately, consensus and conformity in the geosciences, instead of critical curiosity and honest scientific open-mindedness, is largely to blame. Instead of questioning and seeking to understand how Earth behaves, geoscientists fell in line with global warming activists and, I submit, became part of one of the greatest scientific misrepresentations since the heyday of the geocentric universe. From long experience I have observed that when scientists say there are n explanations for a phenomenon, there are usually at least $n+1$ explanations, the 1 being the explanation not yet conceived. I found that to be the case for the composition of the inner parts of the Earth, our planet's primary internal energy sources, origin of the geomagnetic field, origin of fold-mountains, thermonuclear ignition of stars, and galactic luminous star distributions, as I describe in this book. Pertaining to the subject of anthropogenic global warming, I found evidence that the Earth's two primary internal energy sources are potentially variable:

- **Georeactor variability is indicated by the periodicity and synchronicity of lava outpourings in Hawaii and Iceland [14, 15].**
- **Variability of the stored protoplanetary energy of compression, the main driving-force for Earth's decompression, is indicated by the increase in number of earthquakes with magnitude ≥6 [16]. (Figure 10.1)**

Variable heat production by those two energy sources, I posit, causes variation in seawater temperature and, concomitantly, variation in atmospheric carbon dioxide. So, therefore one must consider instead global *non-anthropogenic* climate change [16], which consequently offers a logical, causally related explanation for the 800,000 year ice-core correlation between local temperature and atmospheric CO_2, shown in Figure 10.2 [17]. The correlation shown in Figure 10.2 certainly cannot have resulted from anthropogenic additions of carbon dioxide to the atmosphere; fossil fuel combustion on a large scale is confined to the last 150 years, just a brief moment in the 800,000 year pattern.

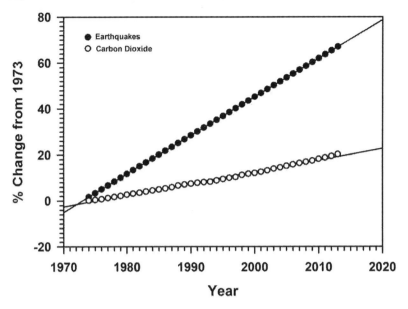

Figure 10.1 (Copy of Figure 1.5) Percent annual change, relative to 1973, of earthquakes, magnitude ≥6, calculated from regression line in Figure 1.1 and from often-quoted CO_2 values from the model-based tabulation published by NASA Goddard Institute for Space Studies (GISS) [18]. From 1981 to 2013, GISS was directed by James E. Hansen, who has a long and public record of being a pro-global warming activist. So, if there is bias, presumably, it is bias on the high-side. This figure shows that earthquakes of magnitude ≥ 6 are increasing at a greater rate than atmospheric CO_2.

Since the first detailed observations of Earth's ocean floor topography, the geoscience community has been locked into the assumption that Earth's heat flux is constant, so little effort was devoted to seeking contrary evidence: That should change. The ice-core correlation between local temperature and atmospheric CO_2 over the entire 800,000 year interval (Figure 10.2) is too consistent to be a fluke and it is certainly not an artifact of human activity. The data cry out for understanding. I have only parted the curtain a bit, proffered a logical and causally related explanation where none had existed before, and provided evidence that Earth's two major energy sources are variable. I demonstrated earthquake variability for the last four decades, and referred to observations of periodicity and synchronicity of lava outpourings over the past 66 million years. While these observations make the point, considerably greater detail is desired and undoubtedly will be discovered. That is the normal progress of science, and exciting opportunities are now available. Essentially, the problem is one of discovering in fine detail the intrinsic interconnections between solar activity, the georeactor (mentioned in Chapter 6), decompression, earthquakes, ocean heating, volcanism, the push-pull of tidal effects, and whatever else that might be contributory. The needs are for scientific discoveries, not making models.

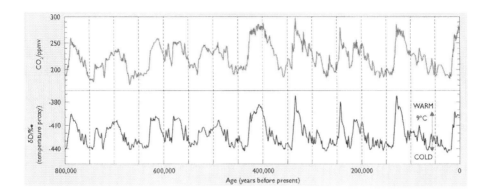

Figure 10.2 (Copy of Figure 1.5) **Correlation between local temperature and trapped CO_2 over a time span of 800,000 years from an Antarctic ice core [17]: Ice core data from the EPICA Dome C (Antarctica) ice core: deuterium is a proxy for local temperature; CO_2 is from the ice core air. Data from [19, 20]. Reproduced with permission of the British Antarctic Survey.**

I wonder if other geophysicists noticed the increase in earthquake frequency that I described, but then failed to report the observation because it could not be explained within their current understanding? If that was indeed the case, it would be a shame. Advances are not made from apparent agreements with extant understanding, but by discovering events that are seemingly inexplicable. There is also a matter of responsibility: The scientific community, especially in earthquake-prone regions, should be made aware so as to advise citizens and governments to prepare.

With understanding and monitoring, the difficulties and dangers that arise as a consequence of variable Earth-heat to some extent can be ameliorated. Attempts at geo-engineering Earth are apt to be unsuccessful as well as disastrous to the creatures of Earth, including our own species.

However, despite these geoscience-based physical threats from the changing Earth that I have identified, a bigger threat – from within our own community – lurks. By far the greatest danger to humanity and to the environment, in my opinion, is not from the inevitable variability of Earth-heat and the ongoing planetary decompression, but from the malevolent political agenda already set in motion. Once set in motion a malevolent political agenda is like an alpine avalanche, virtually unstoppable.

So, what can be done? Are Americans and people of other nations to be blinded of the impending danger by self-optimism, as were many German Jews in the Third Reich, until it is too late? What are the possible futures?

Imagine the future is a hall with a series of closed doors. Behind the first door are the scientific organizations, the American Geophysical Union and the American Physical Society. Under ideal circumstances these might provide the unifying scientific integrity to oppose the avalanche, but for decades scientific integrity seems to have gone missing; the door is jammed. Behind the second door is a Congress eager to undo decades of government malfeasance that led to the avalanche. If that door can be opened, positive, corrective changes will be made, such as replacing the National Science Foundation with an entity that is not underpinned by flawed methodologies that trivialize and corrupt science. But it is not certain that the second door can be opened; it takes courage, strength, and willpower. But if the first two doors cannot be opened, the third door surely can and will be opened. Behind the third door are a host of Pandora's Box-like evils engendered or exacerbated by the malevolent political agenda: human afflictions, ice age onset, famine, environmental devastation, biota poisoning, an increasingly enslaved population, and poised armies seeking retaliation for the real or supposed harm caused to their nations and peoples. There may be still more dangers, but as of now it's too crowded to divine what further perils await behind the third door.

The most serious impending dangers we face, I submit, are not those suggested by a thorough knowledge of the geosciences and their relationship to life – but by our own all-too-human failings. Thus what we need to watch out for is described less in this final chapter than in the first, where, without the scientific background presented in subsequent chapters, I:

- **Laid out my case against the dangerous political hijacking of flawed climate science,**
- **Revealed the increasing occurrence of earthquakes of magnitude ≥ 6, and**
- **Described a new non-anthropogenic basis for atmospheric carbon dioxide increases and for climate change.**

May I humbly suggest that to get the most from this book, the reader might like to re-read the first chapter, now with greater understanding.

Chapter 10 References

1.   Herndon, J.M., *The nickel silicide inner core of the Earth.* Proc. R. Soc. Lond, 1979. A368: p. 495-500.

2.   Birch, F., *The transformation of iron at high pressures, and the problem of the earth's magnetism.* Am. J. Sci., 1940. 238: p. 192-211.

3.   Murrell, M.T. and D.S. Burnett, *Actinide microdistributions in the enstatite meteorites.* Geochim. Cosmochim. Acta, 1982. 46: p. 2453-2460.

4.   API, *AIP Endorsement of American Geophysical Union Climate Change Statement,* 2004.

5.   Herndon, J.M., *Nuclear fission reactors as energy sources for the giant outer planets.* Naturwissenschaften, 1992. 79: p. 7-14.

6.   Herndon, J.M., *Feasibility of a nuclear fission reactor at the center of the Earth as the energy source for the geomagnetic field.* Journal of Geomagnetism and Geoelectricity, 1993. 45: p. 423-437.

7.   Herndon, J.M., *Planetary and protostellar nuclear fission: Implications for planetary change, stellar ignition and dark matter.* Proceedings of the Royal Society of London, 1994. A455: p. 453-461.

8.   Raghavan, R.S., *Detecting a nuclear fission reactor at the center of the earth.* arXiv:hep-ex/0208038, 2002.

9.   Bogue, S.W., *Very rapid geomagnetic field change recorded by the partial remagnetization of a lava flow* Geophys. Res. Lett., 2010. 37: p. doi: 10.1029/2010GL044286.

10.   Coe, R.S. and M. Prevot, *Evidence suggesting extremely rapid field variation during a geomagnetic reversal.* Earth Planet. Sci. Lett., 1989. 92: p. 192-198.

11.   Sagnotti, L., et al., *Extremely rapid directional change during Matuyama-Brunhes geomagnetic polarity reversal.* Geophys. J. Int., 2014. 199: p. 1110-1124.

12. Herndon, J.M., *Nuclear georeactor origin of oceanic basalt ³He/⁴He, evidence, and implications.* Proceedings of the National Academy of Sciences USA, 2003. 100(6): p. 3047-3050.

13. Higgins, P.A.T., *How to deal with climate change.* Physics Today, 2014. 67(10): p. 32.

14. Mjelde, R. and J.I. Faleide, *Variation of Icelandic and Hawaiian magmatism: evidence for co-pulsation of mantle plumes?* Mar. Geophys. Res., 2009. 30: p. 61-72.

15. Mjelde, R., P. Wessel, and D. Müller, *Global pulsations of intraplate magmatism through the Cenozoic.* Lithosphere, 2010. 2(5): p. 361-376.

16. Herndon, J.M., *Variable Earth-heat production as the basis of global non-anthropogenic climate change.* Submitted to Current Science, September 4, 2014.

17. BAS, *Science Briefing - Ice Cores and Climate Change,* http://www.antarctica.ac.uk/press/journalists/resources/science/ice corebriefing.php.

18. NASA_GISS, *http://data.giss.nasa.gov/modelforce/ghgases/Fig1A.ext.txt.*

19. Jouzel, J., et al., *Orbital and millennial Antarctic climate variability over the last 800,000 years.* Sci., 2007. 317: p. 793-796.

20. Lüthi, D., ey al., *High-resolution carbon dioxide concentration record 650,000-800,000 years before present.* Nature, 2008. 453: p. 379-382.

Appendix A

The nickel silicide inner core of the Earth

J. M. Herndon
University of California, San Diego

Published in *Proceedings of the Royal Society of London*, A368, 495-500 (1979)
Communicated by H. C. Urey – Received 27 November 1978 – Revised 19 April 1979

Abstract: From observations of nature, the suggestion is made that Earth's inner core consists, not of nickel-iron metal, but of nickel silicide.

Contemporary understanding of the physical state and chemical composition of the interior of the Earth is derived primarily from interpretations of seismological measurements and from inferences drawn from observations of meteorites. Seismological investigations of Oldham (1906), Gutenberg (1914) and others helped to establish the idea that a fluid core extends to approximately one half the radius of the Earth. The existence of a small, apparently solid inner core at the centre of the Earth was recognized by Lehmann (1936) from interpretations of earthquake records and has been confirmed by Gutenberg and Richter (1938), Jeffreys (1939), Bullen (1957) and others. Observations of meteorites consisting almost entirely of nickeliferous iron led, by inference, to the idea that the fluid core of the Earth consists of molten nickel-iron metal (see, for example, Buddington, 1943; Daly, 1943). Birch (1952) found from density distribution calculations, however, that pure iron (and, to a greater extent, nickel-iron alloy) would be more dense, by 5 - 15%, than the calculated core density, thus indicating that one or more light elements are alloyed with nickel-iron in the core. The variety of combinations of light elements proposed (table 1) can be taken as an indication of contemporary uncertainty in understanding the chemical composition of the interior of the Earth.

Because elements heavier than iron and nickel are less than 1% as abundant in meteorites, it is widely believed that the inner core of the Earth, which has a mass of *ca.* 5% of that of the core, consists of partially crystallized, nickel-iron metal (Brett 1976). Such a metastable state would require the temperature at the inner core boundary to be equal to the melting point of nickel-iron at the relevant pressure and would require the heat content of the core to remain essentially unchanged with the passage of time. Otherwise, the inner core would either grow or diminish. The alternative suggestion that the inner core is the result of a pressure-induced electronic transition in iron (Elsasser & Isenberg 1949; McLachlan & Ehlers 1971) appears, from recent calculations (Bukowinski & Knopoff 1976) to be untenable.

TABLE 1. LIGHT ELEMENTS SUGGESTED BEING ALLOYED WITH NICKEL–IRON
IN THE FLUID CORE OF THE EARTH

C, Si, H (Birch 1952)	Si (MacDonald & Knopoff 1958; Ringwood 1958)
C, S (Urey 1960)	C, S, Si (Clark 1963)
Si, O, S (Birch 1964)	S (Mason 1966; Murthy & Hall 1970; Lewis 1973)
Mg, O (Adler 1966)	O (Bullen 1973; Ringwood 1977)

The Earth and meteorites were derived from primordial matter of common origin. The individual elements that comprise the Earth and meteorites are ensembles of specific nuclear species which occur in remarkably unique relative proportions (figure 1).

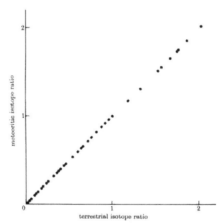

FIGURE 1. Comparison of meteoritic and terrestrial isotope ratios of 15 elements. Some of the 52 isotope ratios plotted from literature values are coincident on this scale (see, for example, Kielbasinski & Wanat 1968).

The relative abundances of many elements in certain types of meteorites, called chondrites, are quite similar to corresponding abundances in the Sun, the latter being obtained from the spectral analysis of sunlight. Primordial elemental abundances appear to be related, although in a complex manner, to nuclear properties (Suess & Urey 1956; Suess & Zeh 1973; Urey 1972).

Physical processes involved in the formation of chondrites did not appreciably separate the less volatile primordial elements from one another (figure 2).

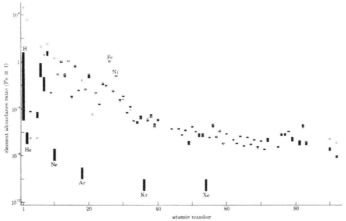

FIGURE 2. Comparison of elemental abundance ratios obtained from spectral analysis of sunlight (○) with corresponding ratios, normalized to iron, obtained from chemical analysis of chondritic meteorites (■): the tops of the bars represent literature values for the hydrated oxygen-rich Orgueil meteorite, the bottoms, those for the anhydrous oxygen-poor Abee enstatite chondrite (see, for example, Holweger 1977).

Certain chondrites, such as the Orgueil meteorite, consist almost exclusively of low temperature hydrated minerals (Bostrom & Fredriksson 1966). Most chondrites, however, contain minerals which formed under anhydrous conditions at temperatures sufficiently high as to have resulted in melting (Rose 1825; Ramdohr 1973). Igneous chondrites differ in their respective oxygen content and, consequently, in the relative proportions of their principle components: silicate/oxide minerals, sulphides and metal (table 2).

TABLE 2. MINERAL ASSEMBLAGES CHARACTERISTIC OF THE CHONDRITIC METEORITES

hydrated chondrites

epsomite, $MgSO_4 . 7H_2O$
complex hydrated layer lattice silicates, e.g. $(Mg, Fe)_6Si_4O_{10}(O, OH)_8$
magnetite, Fe_3O_4

anhydrous chondrites

oxygen-rich carbonaceous chondrites	*ordinary chondrites*	*oxygen-poor enstatite chondrites*
pentlandite, $(Fe, Ni)_9S_8$	troilite, FeS	complex mixed sulphides e.g. $(Ca, Mg, Mn, Fe)S$
troilite, FeS	olivine $(Fe, Mg)_2SiO_4$	pyroxene, $MgSiO_3$
olivine $(Fe, Mg)_2SiO_4$	pyroxene $(Fe, Mg)_2SiO_3$	metal (Fe–Ni–Si alloy)
pyroxene $(Fe, Mg)SiO_3$	metal (Fe–Ni alloy)	nickel silicide, Ni_2Si

Many of the oxygen-rich carbonaceous chondrites contain almost no metal (Herndon, *et al.* 1976). Most chondrites, however, contain less oxygen.

Certain of these, called ordinary chondrites, consist principally of iron oxide-bearing silicate minerals, iron sulphide and nickeliferous iron metal (Keil 1962). The metal of ordinary chondrites is similar in composition to the metal meteorites that evoked early ideas about the core of the Earth. The oxygen-poor enstatite chondrites, on the other hand, consist of silicates almost totally devoid of iron oxide. Iron sulphide occurs, as do sulphides of some elements, such as magnesium, that occur as silicate/oxide minerals in ordinary chondrites. Elemental silicon is present in the metal, and nickel silicide occurs. Meteoritic nickel silicide occurs both as lamellar exsolutions from silicon-bearing iron metal (Ramdohr & Kullerud 1962; Ramdohr 1964; Fredriksson & Henderson 1965; Wasson & Wai 1970; Wai 1970) and as more massive forms intimately associated with metal and iron sulphide in certain enstatite chondrites (Reed 1968; Ramdohr 1973). Among different meteorites, little variation is reported in elemental composition has been reported: 75 - 81% Ni, 3 - 7% Fe, 12 - 15% Si and 2 - 5% P.

At ambient pressure nickel silicide of composition Ni_2Si has a melting point (1309°C) nearly as high as that of pure iron (1533°C) and even higher than the melting points of some iron-based alloys (Hansen 1958). At ambient temperature and pressure, Ni_2Si has a density at of 7.2 g/cm^3; Ni_3Si, 7.9 g/cm^3. The density of nickel silicide is almost identical to that of pure iron (7.86 g/cm^3) and is thus more than the uncompressed density of the core of the Earth (Birch 1952). The occurrence of nickel silicide in enstatite chondrites is proof of its insolubility in oxygen-poor primordial matter.

I suggest that the inner core of the Earth consists of nickel silicide that crystallized from the liquid and settled by the action of gravity to the centre of the Earth. Because nickel is *ca.* 5% as abundant as iron in primordial matter, a fully crystallized inner core of the composition of meteoritic nickel silicide would comprise a mass remarkably similar to that inferred from seismological data. The existence of a nickel silicide inner core would indicate that the interior of the Earth formed from primordial matter that was sufficiently oxygen-poor as to have caused elemental silicon to be present in the iron liquid. The physical state and chemical composition of the Earth, as derived from earthquake waves times of travel and from the above considerations, is illustrated in figure 3.

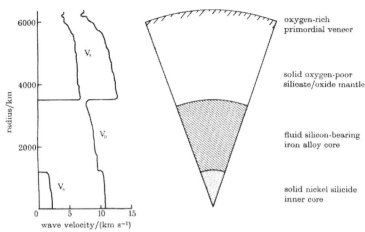

FIGURE 3. The principal divisions and physical state of the interior of the Earth is indicated by the compressional velocity, V_p, and the shear velocity, V_s, of earthquake waves (after Anderson *et al.* 1971). The fluid core cannot support shear waves. The identification of these features with the principal components of oxygen-poor primordial matter is indicated. The oxidized iron and siderophile element content of rocks from the surface regions suggests a veneer of more oxygen-rich primordial matter. The depth and extent of mixing with the reduced silicate/oxide minerals is unknown.

REFERENCES

Alder, B. J. 1966 *J. geophys. Res.* **71**, 4973-4979.

Anderson, D. L. 1971 Sammis, C. and Jordan, T., 1971 *Science, N. Y.* **171**, 1103-1112.

Birch, F. 1952 *J. geophys. Res.* **57**, 227-286.

Birch, F. 1964 *J. geophys. Res.* **69**, 4377-4388.

Boström, *K* & Fredriksson, K. 1966 Smithsonian misc. Collns **151**, No. 3. P. 39.

Brett, R. 1976 *Revs. Geophys. Space Phys.* **14**, 375-383.

Buddington A. F. 1943, *Am. Mineral.* **28**, 119-140.

Bukowinski, M.S.T. & Knopoff, L. 1976 *Geophys. Res. Lett.* **3**, 45-48.

Bullen, K. E. 1957 *Nature, Lond.* **180**, 49-50.

Bullen, K. E. 1973 *Nature, Lond.* **243**, 68-70.

Clark, S. P., Jr. 1963 in *The earth sciences*, (ed. T. W. Donnelly) Chicago: University of Chicago Press.

Daly, R. A. 1943 *Bull. geol. Soc. Am.* **54**, 401-456.

Elsasser, W. M. & Isenberg, I. 1949 *Phys. Rev.* **76**, 469.

Fredriksson, K & Henderson, E. P. 1965 *Trans. Am. geophys. Un.* **46**, 121.

Gutenberg, B. 1914 *Göttinger Nachr.* **166**, 218.

Gutenberg, B. & Richter, C. F. 1938 *Mon. Nat. R. astr. Soc. geoph. Suppl.* **4**, 363-372.

Hansen, M. 1958 *Constitution of binary alloys.* New York: McGraw-Hill.

Herndon, J. M., Rowe, M. W., Larson, E. E. & Watson, D. W. 1974 *Earth Planet. Sci. Lett.* **29**, 283-290.

Holweger, H. 1977 *Earth Planet. Sci. Lett.* **34**, 152-154.

Jeffreys, H. 1939 *Mon. Nat. R. astr. Soc. geophys. Suppl.* **4**, 548-561.

Keil, K. 1962 *J. geophys. Res.* **67**, 4055-4061.

Kielbasinski, J. & Wanat, L. 1968 *Nuclear Energy Information Center Review Report* no. **32**, Warsaw, Poland.

Lehmann, I. 1936 *Publs. Bur. centr. sism. int.* **A14**, 3-31.

Lewis, .J. S. 1973 *A. Rev. phys. chem.* **24**, 339-351.

MacDonald, G.J.F. & Knopoff, L. 1958 *Geophys. J.* **1**, 284-297.

Mason, B. 1966 *Nature, Lond.* **211**, 616-618.

McLachlan, D. W. & Ehlers, E. 1971 *J. geophys. Res.* **76**, 2780-2789.

Murthy, V. R. & Hall, H. T. 1972 *Phys. Earth planet. Interiors* **6**, 123-130.

Oldham, R. D. 1906 *Q. J. geol. Soc. Lond.* **62**, 456 475.

Ramdohr, P. 1964 Sber. dt. Akad. Wiss., Kl. Chem., Geol., Biol., no. 5, 1- 40.

Ramdohr, P. 1973 *The opaque minerals in stony meteorites.* New York: Elsevier.

Ramdohr, P. & Kullerud, G. 1962 *Carnegie Inst. Wash. Yb.* **61**, 163.

Reed, S.J.B. 1968 *Mineralog. Mag.* **36**, 850-854 ..

Ringwood, A. E. 1958 *Geochim. cosmochim. Acta* **15**, 195-212.

Ringwood, A, E. 1977 *Publ.* No. **1299**, Research School of Earth Sciences, Australian National University, Canberra.

Rose, G. 1825 *Annln. Phys.* **4**, 173-204; (abstract) *Edinb. phil. J.* **13**, 368.

Suess, H. E. & Urey, H. C. 1956 *Rev. mod. Phys.* **28**, 53-74.

Suess, H. E. and Zeh, H. D. 1973 *Astrophys. & Space Sci.* **23**, 173-187.

Urey, H. C. 1960 *Geochim. cosmochim. Acta* **18**, 151-153.

Urey, H. C. 1972 *Ann. N.Y. Acad. Sci.* **194**, 35-44.

Wai, C. M. 1970 *Mineral. Mag.* **37**, 905-908.

Wasson, J. T. & Wai, C. M. 1970 *Geochim. cosmochim. Acta* **34**, 169-184.

Appendix B

Terracentric Nuclear Fission Georeactor: Background, Basis, Feasibility, Structure, Evidence, and Geophysical Implications

J. Marvin Herndon
Transdyne Corporation

Published in

Current Science, Vol. 106(4), 25 February 2014, pp. 528-541.
Download PDF http://www.nuclearplanet.com/0528.pdf

Abstract: The background, basis, feasibility, structure, evidence, and geophysical implications of a naturally occurring Terracentric nuclear fission georeactor are reviewed. For a nuclear fission reactor to exist at the center of the Earth, all of the following conditions must be met: (1) There must originally have been a substantial quantity of uranium within Earth's core; (2) There must be a natural mechanism for concentrating the uranium; (3) The isotopic composition of the uranium at the onset of fission must be appropriate to sustain a nuclear fission chain reaction; (4) The reactor must be able to breed a sufficient quantity of fissile nuclides to permit operation over the lifetime of Earth to the present; (5) There must be a natural mechanism for the removal of fission products; (6) There must be a natural mechanism for removing heat from the reactor; (7) There must be a natural mechanism to regulate reactor power level; and; (8) The location of the reactor must be such as to provide containment and prevent meltdown. Herndon's georeactor alone is shown to meet those conditions. Georeactor existence evidence based upon helium measurements and upon antineutrino measurements is described. Geophysical implications discussed include georeactor origin of the geomagnetic field, geomagnetic reversals from intense solar outbursts and severe Earth trauma, as well as georeactor heat contributions to global dynamics. The article is organized as follows: 1.0 Introduction and Background; 2.0 Georeactor Basis; 2.1 Uranium in Earth's Core; 2.2 Uranium Concentration; 2.3 Georeactor Nuclear Calculations; 2.4 Requisite Uranium Isotopic Composition; 2.5 Requisite Fuel Breeding; 2.6

Requisite Fission Product Removal; 2.7 Requisite Georeactor Heat Removal; 2.8 Requisite Regulation Mechanism; 2.9 Requisite Georeactor Containment; 3.0 Georeactor Existence Evidence Based on Helium Measurements; 4.0 Georeactor Existence Evidence Based on Antineutrino Measurements; 5.0 Heat Flow Considerations and Georeactor Contributions to Geodynamics; 6.0 Georeactor Origin of the Geomagnetic Field; 7.0 Georeactor Geomagnetic Reversals; 8.0 Perspectives, followed by Acknowledgements and References.

1.0 Introduction and Background

The 1938 discovery of nuclear fission [1, 2], the splitting of the uranium nucleus, fundamentally changed human perceptions on warfare, energy production and the nature of planet Earth. Much has been written on the subject of nuclear fission weapons and power plants; comparatively little has been written about terrestrial natural nuclear fission. The latter is the subject of this review.

In 1939, Libby attempted to discover whether uranium in nature undergoes spontaneous nuclear fission [3]. His negative result implied that, if uranium could decay by spontaneous fission, the spontaneous fission half-life would be greater than 10^{14} years. In 1940, Flerov and Petrzhak [4, 5] announced the discovery of spontaneous nuclear fission in uranium with a half-life of about 10^{17} years.

During this period and extending to 1950, there was discussion of the possibility of large-scale nuclear reactions in the Earth's crust or mantle [6-8]. In 1953, Fleming and Thode [9], and Wetherill [10] studied the isotopic compositions of krypton and xenon extracted from uranium minerals and discovered that the fissionogenic isotope assemblages could be understood as a binary mixture of components from spontaneous fission and neutron-induced fission. Notably, in samples containing prodigious neutron absorbers, the neutron-induced component was low or absent, whereas in older samples with less neutron absorbers, the relative amount of neutron-induced fission in the mixture was significantly greater. In fact, in one sample of Belgian Congo pitchblend, Wetherill and Ingram [11] reported a 35% neutron-induced fission proportion in that mixture, leading them to state: "Thus the deposit was twenty-five percent of the way to becoming a pile [nuclear fission reactor]. It is also interesting to extrapolate back 2,000 million years where the ^{235}U abundance was 6% instead of 0.7%. Certainly, such a deposit would be closer to being an operating pile".

In 1956, Kuroda [12, 13] applied Fermi's nuclear reactor theory [14] and demonstrated the feasibility that 2,000 million years ago seams of uranium ore 1 m in thickness could engage in neutron-induced nuclear fission chain reactions. He predicted that ground water would serve as a moderator slowing neutrons to thermal energy levels. For 16 years, Kuroda later told me, the subject of natural nuclear fission reactors was unpopular in the geoscience community. In fact, he related, the only way those papers got published at all was that at the time the *Journal of Chemical Physics* would publish short papers without peer-review. But then, reality struck: Kuroda's prediction was proven to have taken place in nature.

In 1972, scientists at the French Commissariat à l' Énergie Atomique announced the discovery of the intact remains of a natural nuclear fission reactor in a uranium mine (Figure 1) at Oklo, near Franceville in the Republic of Gabon in Western Africa. Seams of uranium ore, 1 m in thickness, had undergone sustained nuclear fission chain reactions 1800 million years ago. The reactor had operated at a low power level, 10-100 kW for a period of several thousand years. Control of the chain reaction appears to have been achieved by the reactor boiling-off of ground water, the moderator for slowing neutrons, effectively shutting down the reaction, which restarted when the cooler environment allowed water to return. Although the chain reaction primarily involved slow (thermal) neutron fission, examination of the fission products showed the reactor had functioned to a lesser extend as a fast neutron breeder reactor, producing additional fissile elements from ^{238}U.

The discovery of fossil nuclear reactors at Oklo represented a profound revelation in human thought: Nuclear fission reactors are very much a part of nature, not simply man-made contrivances [15-19]. Significantly, Oklo studies helped to pave the way for the next important advance, my demonstration of the feasibility of a nuclear fission reactor at the centre of the Earth.

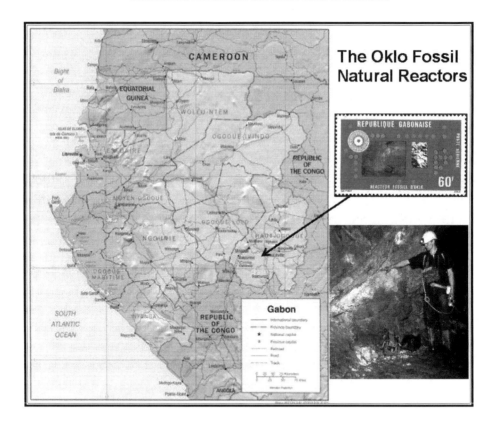

Figure 1. Location of the natural reactors at Oklo, near Franceville, in the Republic of Gabon in Western Africa, indicated by the commemorative postage stamp issued by that nation in honor of those natural reactors. Inset photo of a uranium seam in the reactor zone courtesy of Francoise Gauthier-Lafaye.

The Earth has near or at its centre a powerful energy source that powers the mechanism that generates the geomagnetic field. In 1993, I published a paper where I used an approach similar to that employed by Kuroda [12, 13], demonstrating the feasibility based upon application of Fermi's nuclear reactor theory [14]. But unlike Kuroda, who had knowledge of uranium deposits on Earth's surface, I had to provide justification for significant uranium occurring in the Earth's core and to provide a mechanism in nature for its concentration. In the intervening years, I developed the concept, understanding better the nature, structure, and geophysical consequences which I described in a series of scientific papers [20-28] and books [29-31]. Here I review the subject of Earth's nuclear fission georeactor, which is shown schematically in Figure 2.

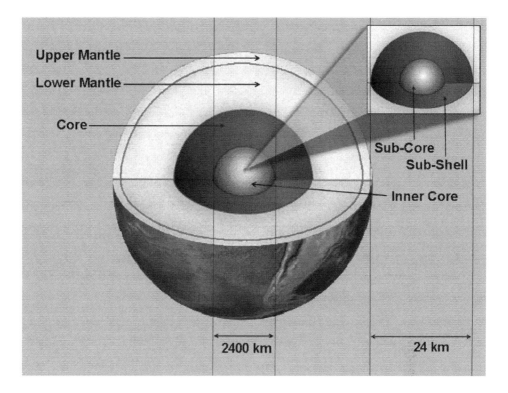

Figure 2. Earth's nuclear fission georeactor (inset) shown in relation to the major parts of Earth. The georeactor at the center is one ten-millionth the mass of Earth's fluid core. The georeactor sub-shell, I posit, is a liquid or a slurry and is situated between the nuclear-fission heat source and inner-core heat sink, assuring stable convection, necessary for sustained geomagnetic field production by convection-driven dynamo action in the georeactor sub-shell [22, 26, 27].

Fundamental concepts from the 1930s and 1940s underpin the current textbook explanation of Earth's structure and composition. Briefly, these are: (i) The Earth resembles an ordinary chondrite meteorite; (ii) the inner core is iron metal in the process of solidifying from the fluid iron alloy core; and, (iii) the silicate mantle is of uniform chemical composition with its observed seismic discontinuities explained as boundaries between pressure-induced changes in crystal structure. Forty years after the inner core's composition was pronounced, I published a different idea for its composition which led me to show, by fundamental ratios of mass, that Earth resembles, not an ordinary chondrite, but a highly reduced enstatite chondrite which implied different deep-Earth chemistry. One chemical consequence is that copious amounts of uranium exist within the core, instead of exclusively in the mantle as previously thought. Subsequently, I demonstrated the feasibility of a Terracentric nuclear fission reactor, and then developed the concept, which is connected to the production of the geomagnetic field, and is a major new energy source with significant geological implications, for example, related to hotspots such as underlies the Hawaiian Islands.

2.0 Georeactor Basis

For a nuclear fission reactor to exist at the centre of the Earth, all of the following conditions must be met:

- There must originally have been a substantial quantity of uranium within Earth's core.
- There must be a natural mechanism for concentrating the uranium.
- The isotopic composition of the uranium at the onset of fission must be appropriate to sustain a nuclear fission chain reaction.
- The reactor must be able to breed a sufficient quantity of fissile nuclides to permit operation over the lifetime of Earth to the present.
- There must be a natural mechanism for the removal of fission products.
- There must be a natural mechanism for removing heat from the reactor.
- There must be a natural mechanism to regulate reactor power level.
- The location of the reactor must be such as to provide containment and prevent meltdown.

In the following subsections, I describe the manner by which each of the above conditions is fulfilled for the Herndon's nuclear fission georeactor at the centre of Earth, and not fulfilled for other, later, putative 'georeactors' assumed to be located elsewhere in Earth's deep interior.

2.1 Uranium in Earth's Core

In 1898, Wiechert suggested that the Earth's mean density could be accounted for if it has a core made of nickeliferous iron metal, like the iron meteorites he had seen in museums [32]. In 1906, Oldham discovered that earthquake-wave velocities increase with depth, but then slow abruptly; he had discovered the Earth's core [33]. In 1936, Lehmann discovered the inner core by reasoning that its existence could account for earthquake waves being reflected into a region where they should otherwise not have been detected [34]. But what is the chemical composition of the inner core?

Studies of earthquake waves and moment of inertia calculations can delineate structures within the Earth and their physical states, but not their chemical compositions. For compositions, one must rely upon implications derived from chondrite meteorites. Chondrite elements, like corresponding elements in the outer portion of the Sun, were never appreciably separated from one another; they thus provide a basis for understanding the bulk composition of Earth. But the situation is complicated because there are three groups of chondrites (ordinary, carbonaceous and enstatite chondrites) that differ significantly in their states of oxidation and in their mineral assemblages [35].

Of the three groups of chondrites, only ordinary chondrites and enstatite chondrites contain appreciable quantities of iron metal. But enstatite chondrites are quite rare and the origin of their highly reduced state of oxidation was not understood. So, ca. 1940, there was widespread perception that the Earth resembles an ordinary chondrite. The inner core was thought to be partially crystallized nickel-iron metal in the process of solidifying by freezing from the Earth's nickel-iron core [36]. That conclusion was reached because in ordinary chondrites nickel is invariably alloyed with iron metal, and elements heavier than nickel are insufficiently abundant to account for a mass as great as the inner core.

Subsequent seismic investigations revealed boundaries just above the core and also in the mantle, 660 km below the Earth's surface, where earthquake waves impinging at an angle change speed and direction. Several such seismic boundaries were also discovered in the upper mantle. Seismic boundaries or discontinuities can potentially arise from two very different causes. Earthquake waves change speed and direction when passing from one substance into a chemically different substance or when traversing the boundary between two different crystal structures of the same material. The former made no sense under the assumption that the Earth is similar to an ordinary chondrite, so pressure-induced crystalline phase boundaries became the explanation that is widely believed even now. But the Earth resembles, not an ordinary chondrite, but an enstatite chondrite and the seismic discontinuities at the depth of 660 km and below characterize the boundaries between different chemical compositions of matter.

Imagine heating a metal-bearing chondrite. At some temperature below the melting points of the silicate minerals, the iron sulfide melds with the nickel-iron metal and forms a liquid capable of percolating downward by gravity forming a two-component system analogous to the structure of Earth. Figure 3 shows that only enstatite chondrites, not ordinary chondrites, harbour a sufficient proportion of iron alloy to account for the massive core of the Earth.

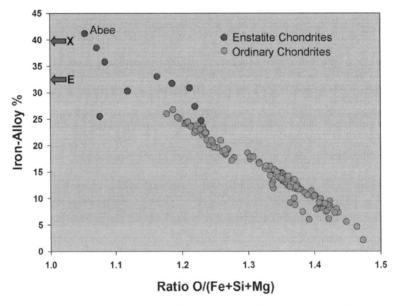

Figure 3. Evidence that Earth resembles an enstatite chondrite. The percent alloy (iron metal plus iron sulfide) of 157 ordinary chondrites (green circles) and 9 enstatite chondrites (red circles) plotted against oxygen content. The core percent of the whole-Earth, "arrow E", and of (core-plus-lower mantle), "arrow X", shows that Earth resembles an Abee-type enstatite chondrite and does not resemble an ordinary chondrite. Data from references [37-40].

Forty years after Birch explained the inner core's composition as being partially crystallized nickel-iron metal, I deduced its composition as fully crystallized nickel silicide [41] based upon discoveries made in the 1960s [42-44]. Subsequently, I discovered that the mass ratios of the components of the inner 82% of the Earth are virtually identical to corresponding components of a primitive enstatite chondrite, as shown in Table 1. That identity means that the components of a primitive enstatite chondrite are compositionally similar to corresponding components in Earth's deep interior. Moreover, it means the deep interior of the Earth has the same highly reduced state of oxidation as a primitive enstatite chondrite. Furthermore, it means that a substantial quantity of uranium occurs in the Earth's core as most, if not all, of the uranium in the primitive Abee enstatite chondrite occurs in the part that corresponds to the Earth's core [45].

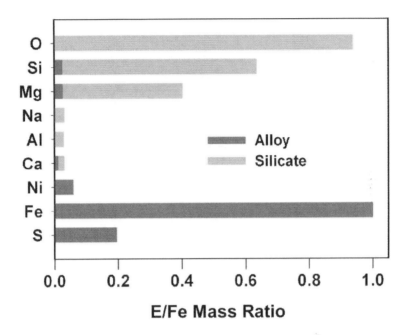

Figure 4. Relative abundances of the major and minor elements in the Abee enstatite chondrite, normalized to iron, showing their relative amounts in the alloy and silicate portions. Note that calcium (Ca), magnesium (Mg), and silicon (Si), normally lithophile elements, occur in part in the alloy portion.

Table 1. Fundamental mass ratio comparison between the endo-Earth (lower mantle plus core) and the Abee enstatite chondrite. Above a depth of 660 km, seismic data indicate layers suggestive of veneer, possibly formed by the late addition of more oxidized chondrite and cometary matter, whose compositions cannot be specified with certainty at this time.

Fundamental Earth Ratio	Earth Ratio Value	Abee Ratio Value
lower mantle mass to total core mass	1.49	1.43
inner core mass to total core mass	0.052	theoretical 0.052 if Ni_3Si 0.057 if Ni_2Si
inner core mass to lower mantle + total core mass	0.021	0.021
D'' mass to total core mass	0.09[***]	0.11[*]
ULVZ[**] of D'' CaS mass to total core mass	0.012[****]	0.012[*]

[*] = avg. of Abee, Indarch, and Adhi-Kot enstatite chondrites
D'' is the "seismically rough" region between the fluid core and lower mantle
[**] ULVZ is the "Ultra Low Velocity Zone" of D''
[***] calculated assuming average thickness of 200 km
[****] calculated assuming average thickness of 28 km
data from [46-48]

2.2 Uranium Concentration

Only five elements comprise about 95% of the mass of a chondrite meteorite and by inference the Earth; the four minor elements add about 3%. Figure 4 shows the distribution of those elements between the alloy and the silicate portions of a primitive enstatite chondrite, and, by the identity shown in Table 1, between the core and mantle of Earth. As consequence of the high state of reduction, certain elements that have a high affinity for oxygen, including calcium, magnesium and silicon, occur in part in the iron alloy portion. The high state of reduction, a consequence of the Earth's formation [25, 27, 49, 50], is the reason that uranium occurs in the Earth's core.

Elements that have a high affinity for oxygen are generally incompatible in an iron alloy. Upon cooling from a high temperature these oxyphile elements escape the iron alloy by precipitating when thermodynamically possible. In the Earth's core, calcium and magnesium reacted with sulfur at a high temperature to form CaS and MgS, which floated to the top of the core. Silicon precipitated by combining with nickel. The nickel silicide sank by the action of gravity and formed the inner core.

In the deep interior of the Earth, density is a function almost exclusively of atomic number and atomic mass. The core is layered on the basis of density. Uranium, more dense than the inner core by more than a factor of two, either as metal or mono-sulfide, driven by gravity, perhaps through a series of steps, concentrated at the gravitational center of Earth. Figure 5, a schematic representation of the deep interior of Earth, illustrates the layering by density which, with the data from Table 1, explains well the observed seismic boundaries as compositional boundaries, not pressure-induced crystalline phase boundaries.

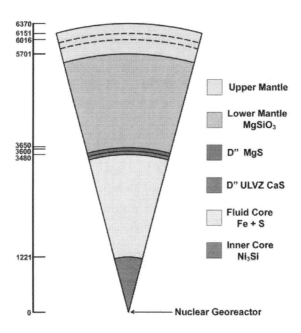

Figure 5. Schematic representation of the layers of the Earth based upon the data from which Table 1 is derived. Scale in km.

2.3 Georeactor Nuclear Calculations

In his nuclear reactor theory, Fermi [14] defined the condition for a self-sustaining nuclear fission chain reaction. The value of k_{eff} represents the number of fission neutrons in the present population divided by the number of fission neutrons in the previous population. The defining condition for self-sustaining nuclear fission chain reactions is that k_{eff} = 1.0. If k_{eff} > 1.0, the neutron population and the energy output are increasing and will continue until changes in the fuel, moderators, and neutron absorbers cause k_{eff} to decrease to 1.0. If k_{eff} < 1.0, the neutron population and energy output are decreasing and will eventually decrease to 0. If k_{eff} = 1.0, the neutron population and energy output are constant.

Initially, I demonstrated the feasibility of a Terracentric nuclear fission georeactor by applying Fermi's nuclear reactor theory. Subsequently, far more sophisticated numerical simulation calculations were made using the SAS2 analysis sequence contained in the SCALE Code Package from Oak Ridge National Laboratory [51] that has been developed over a period of three decades and has been extensively validated against isotopic analyses of commercial reactor fuels [52-56]. The SAS2 sequence invokes the ORIGEN-S isotopic generation and depletion code to calculate concentrations of actinides, fission products, and activation products simultaneously generated through fission, neutron absorption, and radioactive decay. The SAS2 sequence performs the 1-D transport analyses at selected time intervals, calculating an energy flux spectrum, updating the time-dependent weighted cross-sections for the depletion analysis, and calculating the neutron multiplication of the system.

The difference between calculations based upon Fermi's nuclear reactor theory and the numerical simulation codes developed at Oak Ridge National Laboratory is similar to the difference between a photograph and a full-length motion picture. For a given configuration of fissionable elements, Fermi's theory allows only determination of the feasibility of a nuclear fission chain reaction at a single moment in time. The Oak Ridge simulations, on the other hand, progress through time in a series of discrete steps. At each step the nuclear reaction-induced changes in fuel composition, which affect the next step, are calculated. These include neutron-induced nuclear reactions, production of fission products, and the natural radioactive decay of fuel and fission products.

The Oak Ridge simulations were a major advance for georeactor calculations because they demonstrated that the georeactor could function over the entire age of the Earth through natural fuel-breeding nuclear reactions. The simulations yield quantitative estimates of specific fission products, one being helium, which serves as a geochemical tracer and offers an explanation for exhaled deep-Earth helium.

2.4 Requisite Uranium Isotopic Composition

Natural uranium consists mainly of the readily-fissionable ^{235}U and the essentially non-fissionable ^{238}U. In a natural nuclear fission reactor, the value of k_{eff} is strongly dependent upon the ratio $^{235}U/^{238}U$. As shown by curves A, B and C in Figure 6, a substantial mass of natural uranium (e.g., a few hundred kilograms) during the first two gigayears of Earth's existence would be capable of undergoing sustained nuclear fission chain reactions. After that point in time, other conditions must also be fulfilled, namely fuel breeding and fission product removal.

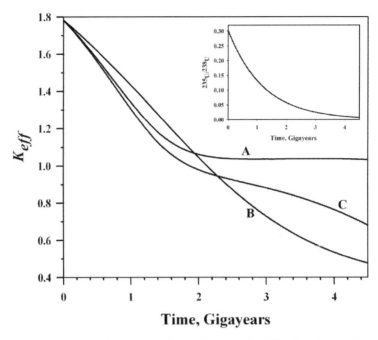

Figure 6. Numerical simulation results, chosen to illustrate main georeactor operational parameters and uncertainties, are presented in terms of k_{eff} over the lifetime of the Earth. The curve labeled A is a 3 TW run in which fuel breeding occurs for the case of fission products instantly removed upon formation. Curve B is the same as A, except fission products are left in place. Curve C is a very low-power run, fission products instantly removed, where the low-level of breeding is insufficient for maintaining criticality. These show the importance of breeding, fission-product removal, and intrinsic self-regulation. Inset shows the natural decay of non-fissioning uranium over the lifetime of Earth.

2.5 Requisite Fuel Breeding

The half-life of [235]U is shorter than that of [238]U. So, radioactive decay causes the ratio [235]U/[238]U to decrease over time, as shown by the inset in Figure 6. Curves B and C in the same figure show that at some point during the decline, neutron absorption by the increasingly greater relative proportion of [238]U causes $k_{eff} < 1.0$, essentially killing the nuclear fission chain reaction, but curve A maintains criticality, i.e., $k_{eff} > 1.0$, by fuel breeding reactions throughout the lifetime of the Earth.

Under appropriate operating conditions, the neutrons produced by nuclear fission can "breed" additional fuel from essentially non-fissionable

^{238}U. As noted in the literature [57] [20], the principal georeactor fuel-breeding takes place by the reaction

$$^{238}\text{U}(n,\gamma)^{239}\text{U}(\beta^-)^{239}\text{Np}(\beta^-)^{239}\text{Pu}(\alpha)^{235}\text{U}.$$

Curve B in Figure 6, calculated at a very low power level, shows the decrease in k_{eff} that would result with too little fuel breeding. Curves A and B were each calculated with fission products promptly removed from the reactor zone.

2.6 Requisite Fission Product Removal

During the first two gigayears of Earth's existence, a deep-Earth nuclear fission georeactor could function without fuel breeding and without removal of fission products. But to maintain the nuclear fission chain reaction into the present requires both fuel breeding and prompt removal of fission products. Even with fuel breeding, k_{eff} will diminish unless fission products are promptly removed. The calculations for curves A and C of Figure 6 were identical, except that fission products were left in place for curve C.

There is a natural process for the removal of fission fragments from a nuclear fission georeactor operating at the centre of the Earth. In the deep interior of the Earth, density is a function almost entirely of atomic number and atomic mass. When the uranium nucleus fissions, it usually splits into two roughly equal parts, each part having approximately half the atomic number and half the atomic mass of the parent uranium. That means that the density of the fission fragments is less than the density of uranium. Consequently, the fission fragments tend to migrate outward and away from the reactor zone, while the uranium tends to re-concentrate downward.

2.7 Requisite Georeactor Heat Removal

The Terracentric georeactor produces heat through nuclear fission and through the decay of radionuclides. There is a natural process for heat removal from the georeactor. The georeactor is thought to have a two-part structure as shown schematically in Figure 7. The concentration of uranium at the centre of the Earth forms the nuclear fission reactor zone, called the georeactor sub-core. This is where the heat is primarily generated by nuclear fission and by the natural decay of actinide elements. Surrounding the sub-core is a shell composed of fission products and the products of radioactive decay, which is referred to as the georeactor sub-shell. Heat is also produced in the georeactor sub-shell by fission-product radioactive decay, although less heat than in the sub-core. The sub-shell is thought to be a liquid or slurry that is engaged in thermal convection [22, 26, 27].

The natural configuration of the georeactor is ideal for heat transport by thermal convection. Heat is produced primarily in the georeactor's nuclear sub-core, which heats the matter at the base of the georeactor's nuclear waste sub-shell causing it to expand, becoming less dense. The less dense 'parcel' of bottom matter floats to the top of the sub-shell where it contacts the massive inner core heat-sink and loses its extra heat, densifies, and sinks. The inner core heat-sink is surrounded by an even more massive heat-sink, the fluid iron alloy core, which helps to ensure the existence of an adverse temperature gradient in the georeactor sub-shell, a necessary condition for thermal convection.

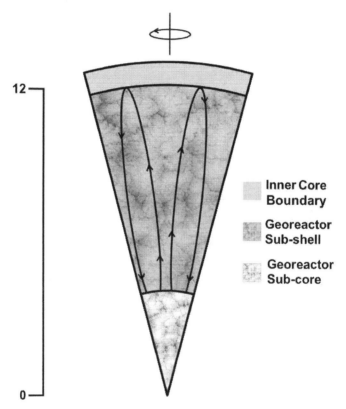

Figure 7. Schematic representation of the georeactor. Planetary rotation and fluid motions are indicated separately; their resultant motion is not shown. Stable convection with adverse temperature gradient and heat removal is expected. Scale in km.

2.8 Requisite Regulation Mechanism

During the first two gigayears of Earth's existence, as indicated in Figure 6, the $^{235}U/^{238}U$ ratio was quite large; initially the readily fissionable ^{235}U comprised 25% of the uranium. So, a highly energetic georeactor was possible in principle, unless the natural configuration of the georeactor affords a self-regulation mechanism.

Georeactor numerical simulations undertaken by Oak Ridge National Laboratory and by Transdyne Corporation were made at constant power levels. Assuming that the georeactor consisted of the maximum amount of available uranium [45], we determined a maximum constant power level of 30 TW [58]. At a higher power level, the georeactor would have fully consumed its fuel and ceased to operate before now. Similarly, too low a power level would be insufficient for the fuel breeding that is necessary for sustained georeactor operation during the last two gigayears (Figure 6, curve B). There must exist a natural georeactor self-regulation mechanism.

In the micro-gravity environment at the centre of the Earth (Figure 7), georeactor heat production that is too energetic would be expected to cause actinide sub-core disassembly, mixing actinide elements with neutron-absorbers of the nuclear waste sub-shell, quenching the nuclear fission chain reaction. But as actinide elements begin to settle out of the mix, the chain reaction would restart, ultimately establishing a balance, a dynamic equilibrium between heat production and actinide settling-out, a self-regulation control mechanism [27].

2.9 Requisite Georeactor Containment

The dense nuclear sub-core of the georeactor, described above, requires containment which is provided naturally by its location at the centre of the Earth. Heat produced by nuclear fission chain reactions cannot cause meltdown as the sub-core already resides at the gravitational bottom. In the wake of interest stimulated by Herndon's georeactor, several attempts were made to describe 'georeactors' at locations other than the centre of the Earth, namely, at the core mantle boundary [59] and atop the inner core [60, 61]. But in each case there is no confinement. If nuclear fission chain reactions were to have occurred in those places, meltdown would inevitably take place and those putative 'georeactors' would meltdown to Earth's center, the location of Herndon's georeactor.

3.0 Georeactor Existence Evidence Based on Helium Measurements

When a uranium nucleus undergoes nuclear fission, it usually splits into two roughly equal, large fragments. Once in every 10,000 fission events, however, the nucleus splits into three pieces, two large and one very small. Tritium, ^{3}H, is a prominent very small fragment of ternary fission. Tritium is radioactive with a half-life of 12.32 years and decays to ^{3}He; ^{4}He is likewise georeactor-produced and also derives from the alpha particles of natural decay. Figure 8 presents helium fission product results from georeactor numerical simulations conducted at Oak Ridge National Laboratory, expressed as ^{3}He/^{4}He relative to the same ratio measured in air [24]. That georeactor-produced ^{3}He/^{4}He ratios have the same range of values observed in oceanic lava is strong evidence that the georeactor exists and is the source of the observed deep-Earth helium [62].

In 1969, Clarke *et al.* [63] discovered that ^{3}He and ^{4}He are venting from the Earth's interior. At the time there was no known deep-Earth mechanism that could account for the experimentally measured ^{3}He, so its ad hoc origin was assumed to be a primordial ^{3}He component, trapped at the time of the Earth's formation, which was subsequently diluted with the appropriate amount of ^{4}He from radioactive decay.

The ^{3}He/^{4}He ratio of helium occluded in basalt at mid-ocean ridges is 8.6 ± 1 times greater than the same ratio in air, expressed as 8.6 R_A. Table 2 shows the commonality of normalized ^{3}He/^{4}He ratios in the 2σ confidence interval for helium trapped in basalt extruded from undersea volcanoes.

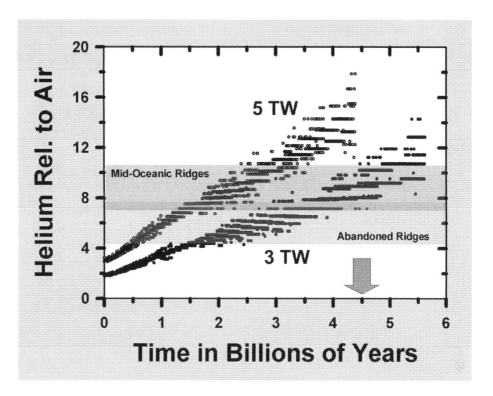

Figure 8. Fission product ratio ^{3}He/^{4}He, relative to that of air, R$_A$, from nuclear georeactor numerical calculations at 5 terawatts, TW, (upper) and 3 TW (lower) power levels [24]. The band for measured values from mid-oceanic ridge basalts is indicated by the solid lines. The age of the Earth is marked by the arrow. Note the distribution of calculated values at 4.5 billion years, the approximate age of the Earth. The increasing values are the consequence of uranium fuel burn-up. Icelandic deep-Earth basalts present values ranging as high as 37 times the atmospheric value [64].

Table 2. Statistics of ^{3}He/^{4}He relative to air (R_A) of basalts from along the global spreading ridge system at a two standard deviation (2σ) confidence level. Adapted from reference [65].

Submarine Basalt Province	^{3}He/^{4}He Relative to Air (R_A)
Propagating Lithospheric Tears	11.75 ± 5.13 R_A
Manus Basin	10.67 ± 3.36 R_A
New Rifts	10.01 ± 4.67 R_A
Continental Rifts or Narrow Oceans	9.93 ± 5.18 R_A
South Atlantic Seamounts	9.77 ± 1.40 R_A
MORB	8.58 ± 1.81 R_A
EM Islands	7.89 ± 3.63 R_A
North Chile Rise	7.78 ± 0.24 R_A
Ridge Abandoned Islands	7.10 ± 2.44 R_A
South Chile Rise	6.88 ± 1.72 R_A
Central Atlantic Islands	6.65 ± 1.28 R_A
HIMU Islands	6.38 ± 0.94 R_A
Abandoned Ridges	6.08 ± 1.80 R_A

The average helium ratios, shown in Table 2, are only part of the overall picture; occluded helium measured in basalt from 'hotspots', such as Iceland and the Hawaiian Islands, typically have ^{3}He/^{4}He ratios relative to air that are greater than 10 times the value in air with some samples measured as high as high as 37 R_A [64].

Numerical simulations of georeactor operation, conducted at Oak Ridge National Laboratory, provide compelling evidence for georeactor existence: georeactor helium fission products matched quite precisely the ^{3}He/^{4}He ratios, relative to air, observed in oceanic basalt as shown in Figure 8. Note in that figure the progressive rise in ^{3}He/^{4}He ratios over time as uranium fuel is consumed by nuclear fission and radioactive decay. The high ^{3}He/^{4}He ratios observed in samples from 'hotspots' are consistent with the sharp increases observed from georeactor simulations as the uranium fuel becomes depleted and ^{4}He diminishes.

Thermal structures, sometimes called mantle plumes, beneath the Hawaiian Islands and Iceland, two high ^{3}He/^{4}He hot-spots, as imaged by seismic tomography [66, 67], extend to the interface of the core and lower mantle, further reinforcing their georeactor-heat origin. The high ^{3}He/^{4}He ratios measured in 'hotspot' lavas appear to be the signature of 'recent' georeactor-produced heat and helium, where 'recent' may extend several hundred million years into the past. Recently, Mjelde and Faleide [68] discovered a periodicity and synchronicity through the Cenozoic in lava outpourings from Iceland and the Hawaiian Islands, 'hotspots' on opposite sides of the globe, that Mjelde *et al.* [69] suggest may arise from variable georeactor heat-production.

4.0 Georeactor Existence Evidence Based on Antineutrino Measurements

As early as 1930, it seemed that energy mysteriously disappeared during the process of radioactive beta decay. To preserve the idea that energy is neither created nor destroyed, 'invisible' particles were postulated to be the agents responsible for carrying energy away unseen. Finally, in 1956 these 'invisible' antineutrinos from the Hanford nuclear reactor were detected experimentally [70].

As early as the 1960s, there was discussion of antineutrinos being produced during the decay of radioactive elements in the Earth. In 1998, Raghavan *et al.* [71] were instrumental in demonstrating the feasibility of their detection. In 2002, Raghavan [72] authored a paper, entitled "Detecting a Nuclear Fission Reactor at the Center of the Earth" wherein he showed that antineutrinos resulting from nuclear fission products would have a different energy spectrum than those resulting from the natural radioactive decay of uranium and thorium. This paper stimulated intense interest worldwide, especially with groups in Italy, Japan and Russia. Russian scientists [73] expressed well the importance: "Herndon's idea about georeactor located at the centre of the Earth, if validated, will open a new era in planetary physics".

The georeactor is too small to be presently resolved from seismic data. Oceanic basalt helium data, however, provide strong evidence for the georeactor's existence [24, 62] and antineutrino measurements have not refuted its existence [74, 75]. To date, detectors at Kamioka, Japan and at Gran Sasso, Italy have detected antineutrinos coming from within the Earth. After years of data-taking, an upper limit on the georeactor nuclear fission contribution was determined to be either 26% (Kamioka) [75] or 15% (Gran Sasso) [74] of the total energy output of uranium and thorium, estimated from deep-Earth antineutrino measurements (Table 3). The actual total georeactor contribution may be somewhat greater, though, as some georeactor energy comes from natural decay as well as from nuclear fission.

Table 3. Antineutrino determinations of radiogenic heat production [74, 75] shown for comparison with Earth's heat loss to space [76]. See original report for discussion and error estimates.

Heat (terawatts)	Source
44.2 TW	global heat loss to space
20.0 TW	antineutrino contribution from ^{238}U, ^{232}Th, and georeactor fission
5.2 TW	georeactor KamLAND data
3.0 TW	georeactor Borexino data
4.0 TW	^{40}K theoretical
20.2 TW	loss to space minus radiogenic
15.0 TW	natural radioactivity of ^{238}U, ^{232}Th, ^{40}K, without nuclear fission

5.0 Heat Flow Considerations and Georeactor Contributions to Geodynamics

The 1940s concepts about Earth's interior are still evident in current assumption based models: the inner core is still assumed to be partially crystallized iron metal, and the mantle is still assumed to be of uniform composition and considerably more oxidized than an enstatite chondrite. Mantle convection was added with the introduction of seafloor spreading and plate tectonics. Within that framework, computational models of "primitive mantle composition" continue to be promulgated. For example, the bulk silicate Earth (BSE) model attempts to model the unknown "primitive mantle composition" from carbonaceous chondrite element abundances, modified by inferences derived from the assumption that observed peridotite compositions result from partial melting of the assumed primitive mantle [77].

In BSE, uranium and thorium are assumed to reside entirely in the crust and mantle. Heat from the radioactive decay of ^{238}U, ^{232}Th and ^{40}K is assumed to be delivered efficiently to the surface by mantle convection. But even if rapid delivery were possible, heat from natural radioactive decay without georeactor fission, ~15 TW, is insufficient to account for Earth's radiated energy [76] (Table 3), and mantle convection is problematic, both from the standpoint of heat transport and geodynamics [49, 78].

The Earth's mantle is bottom heavy, 62% denser at the bottom than at the top [46]. The small amount of thermal expansion at the bottom (<1%) cannot overcome the 62% higher density at the bottom of the mantle; bottom mantle matter cannot float to the mantle top. Sometimes attempts are made to obviate the 'bottom heavy' prohibition by adopting the tacit assumption that the mantle behaves as an ideal gas, with no viscous losses, i.e., 'adiabatic'. But the mantle is a solid that does not behave as an ideal gas as demonstrated by earthquakes occurring at depths as great as 660 km. Earthquakes in the upper mantle indicate the catastrophic release of stress, observations inconsistent with adiabatic absence of viscous loss.

Mantle silicates are thermal insulators. Absent mantle convection, heat flow to the surface may be extremely inefficient. Moreover, without mantle convection, plate tectonics lacks a valid scientific basis. But that should not be surprising as there are other problems. For example, nowhere in the literature of plate tectonics is presented a logical, causally related explanation for the fact that approximately about 41% of the Earth's surface is continental rock (sial) with the balance being ocean floor basalt (sima). Without mantle convection, there is no motive force for continuing sequences of continent collisions, thought to be the sole basis for fold-mountain formation. The reasonable conclusion therefore is that there must exist a new and fundamentally different geoscience paradigm which obviates the problems inherent in plate tectonics and in planetesimal Earth formation, and yet is capable of better explaining observed geological features.

I have disclosed a new indivisible geoscience paradigm, called Whole-Earth Decompression Dynamics (WEDD) [79], that begins with and is the consequence of our planet's early formation as a Jupiter-like gas giant (Figures 9 and 10) and which permits deduction of: (i) Earth's internal composition and highly-reduced oxidation state (Figure 5); (ii) core formation without whole-planet melting; (iii) powerful new internal energy sources, protoplanetary energy of compression and georeactor nuclear fission energy; (iv) mechanism for heat emplacement at the base of the crust

[80]; (v) nuclear fission georeactor geomagnetic field generation [26]; (vi) decompression-driven geodynamics that accounts for the myriad of observations attributed to plate tectonics without requiring physically-impossible mantle convection [30, 81]; and, (vii) a mechanism for fold-mountain formation that does not necessarily require plate collision [82] (Figure 11).

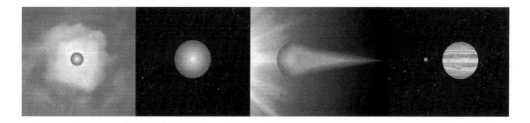

Figure 9. Whole-Earth Decompression Dynamics formation of Earth. From left to right, same scale: 1) Earth condensed at the center of its giant gaseous protoplanet; 2) Earth, a fully condensed a gas-giant; 3) Earth's primordial gases stripped away by the Sun's T-Tauri solar eruptions; 4) Earth at the onset of the Hadean eon, compressed to 66% of present diameter; 5) Jupiter for size comparison.

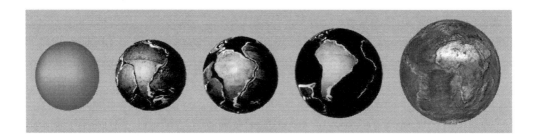

Figure 10. Schematic representation of the decompression of Earth (WEDD) from Hadean to present. From left to right, same scale: 1) Earth after T-Tauri removal of gases, 66% of present Earth diameter, fully covered with continental-rock crust; 2), 3), and 4) Formation of primary and secondary decompression cracks that progressively fractured the continental crust and opened ocean basins. Timescale not precisely established. 5) Holocene Earth.

The geology of Earth, according to WEDD, is principally determined by Earth's decompression: Surface crack formation to accommodate increased planetary volume, and mountain formation to accommodate changes in surface curvature (Figure 11).

Figure 11. Demonstration illustrating the formation of fold-mountains as a consequence of Earth's early formation as a Jupiter-like gas giant. On the left, two balls representing the relative proportions of 'present' Earth (pink), and 'ancient' Earth (blue) before decompression. In the center, a spherical section, representing a continent, cut from 'ancient' Earth and placed on the 'present' Earth, showing: (1) the curvature of the 'ancient continent' does not match the curvature of the 'present' Earth and (2) the 'ancient continent' has 'extra' surface area confined within its fixed perimeter. On the right, tucks remove 'extra' surface area and illustrate the process of fold-mountain formation that is necessary for the 'ancient' continent to conform to the curvature of the 'present' Earth. Unlike the ball-material, rock is brittle so tucks in the Earth's crust would break and fall over upon themselves producing fold-mountains.

Heat from georeactor nuclear fission and radioactive decay can augment mantle decompression by replacing the lost heat of protoplanetary compression. The resulting decompression will tend to propagate throughout the mantle, like a tsunami, until it reaches the barrier posed by the base of the crust. There, crustal rigidity opposes continued decompression, pressure builds and compresses matter at the mantle-crust interface, resulting in compression heating, which, I posit, is the origin the heat responsible for the geothermal gradient [80].

Georeactor-produced heat, channeled to Earth's surface [49], provides a conduit through which highly mobile helium can readily migrate upward. Ultimately, the helium is occluded in volcanic lava produced by that heat. Seismically-observed heat channeling, beneath Iceland and the Hawaiian Islands, extends downward to the top of the core [66, 67]. The association of these 'hotspots' with helium, characterized by $^3He/^4He > 10\ R_A$, is consistent with the georeactor origin of that heat. Generalizing, high helium isotope ratios, $^3He/^4He > 10\ R_A$, may be taken as a signature of georeactor-heat in geological circumstances, even those in which the deep-extending seismic low-velocity profile is no longer evident, e.g., the Siberian traps, which formed 250 million years ago [83].

Some mantle plumes, characterized occluded $^3He/^4He > 10\ R_A$, appear to be associated with continent fragmentation [84]. Georeactor-produced heat is currently involved in the decompression-driven continent-splitting currently occurring along the East African Rift System and may be associated with the petroleum and natural gas deposits discovered there and in other similar circumstances [85]. Generalizing, georeactor-produced heat may augment decompression-driven rift formation that either splits continents or fails to split continents. A more lengthy discussion is available in the literature. [49, 50, 78, 85].

6.0 Georeactor Origin of the Geomagnetic Field

In 1939, Elsasser began a series of scientific publications in which he proposed that the geomagnetic field is produced within the Earth's fluid core by an electric generator mechanism, also called a dynamo mechanism [86-88]. He proposed that convection currents in the Earth's electrically-conducting iron-alloy core, twisted by Earth's rotation, act like a self-sustaining dynamo, a magnetic amplifier, producing the geomagnetic field. For decades, Elsasser's dynamo-in-the-core has been generally considered to be the only potentially viable means for producing the geomagnetic field and has been generally accepted without question. But, I discovered, there are serious problems, not with his idea of a convection-driven dynamo, but with its location within the Earth and with its energy source [26, 27, 49, 50].

There are periods in geological history when the geomagnetic field has been stable for millions of years. Such long-term convection stability cannot be expected in the Earth's fluid core. Not only is the bottom-heavy core about 23% denser at the bottom than the core-top, but the core is surrounded by a thermally insulating blanket, the mantle, with lower thermal conductivity and lower heat capacity than the core. Heat brought to the top of the core cannot be efficiently removed; the core top cannot be maintained at a lower temperature than the core bottom as necessary for thermal convection. Moreover, there is no obvious source of magnetic seed fields in the fluid core.

I have suggested that the geomagnetic field is produced by Elsasser's convection-driven dynamo operating within nuclear waste sub-shell of the georeactor [26]. Unlike the Earth's core, sustained convection appears to be quite feasible in the georeactor sub-shell (Figure 7). The top of the georeactor sub-shell is in contact with the inner core, a massive heat sink, which is in contact with the fluid core, another massive heat sink. Heat brought from the nuclear sub-core to the top of the georeactor sub-shell by convection is efficiently removed by these massive heat sinks thus maintaining the sub-shell adverse temperature gradient. Moreover, the sub-shell is not bottom heavy. Unlike the fluid core, decay of neutron-rich nuclides in the nuclear waste sub-shell provides electrons that might form the seed magnetic fields for amplification.

7.0 Georeactor Geomagnetic Reversals

From time to time, on an irregular basis, the geomagnetic field reverses; north becomes south and vice versa. The average time between reversals is about 250,000 years; the last reversal of the geomagnetic field occurred about 750,000 years ago. There are, however, periods of time up to 50 million years in length when the Earth's magnetic field did not reverse at all.

Reversals of the geomagnetic field are produced when stable convection is interrupted in the region where convection-driven dynamo action occurs, in the nuclear waste sub-shell of the georeactor. Upon re-establishing stable convection, convection-driven dynamo action resumes with the geomagnetic field either in the same or in the reversed direction. The mass of the georeactor is quite low, less than one ten-millionth the mass of fluid core. Consequently, reversals can occur much more quickly, and with greater ease, than previously thought.

Trauma to the Earth, such as a massive asteroid impact or the violent splitting apart of continental land masses, might de-stabilize georeactor dynamo-convection, causing a magnetic reversal or excursion. It is also possible that a geomagnetic reversal might be caused by a particularly violent event on the Sun.

The Earth is constantly bombarded by the solar wind, a fully ionized and electrically conducting plasma, heated to about 1 million degrees Celsius, that streams outward from the Sun and assaults the Earth at a speed of about 1.6 million km/h. The geomagnetic field deflects the brunt of the solar wind safely past the Earth, but some charged particles are trapped in donut-shaped belts around the Earth, called the Van Allen Belts. The charged particles within the Van Allen Belts form a powerful ring current that produces a magnetic field that opposes the geomagnetic field near the equator. If the solar wind is constant, then the ring current is constant and no electric currents are transferred through the magnetic field into the georeactor by Faraday's induction. High-intensity changing outbursts of solar wind, on the other hand, will induce electric currents into the georeactor, causing ohmic heating in the sub-shell, which in extreme cases might disrupt convection-driven dynamo action and lead to a magnetic reversal or excursion.

From ancient lava flows, scientists have recently confirmed evidence of episodes of rapid geomagnetic field change – six degrees per day during one reversal and another of one degree per week – were reported [89, 90]. The relatively small mass of the georeactor is consistent with the possibility of a magnetic reversal occurring on a time scale as short as one month or several years.

Nuclear fission consumes uranium at a much faster rate than natural radioactive decay. At some unknown time in the future, disruption of convection-driven dynamo action will occur, but unlike a magnetic reversal or excursion, there will be insufficient uranium remaining to re-establish stable convection and stable geomagnetic field production [24].

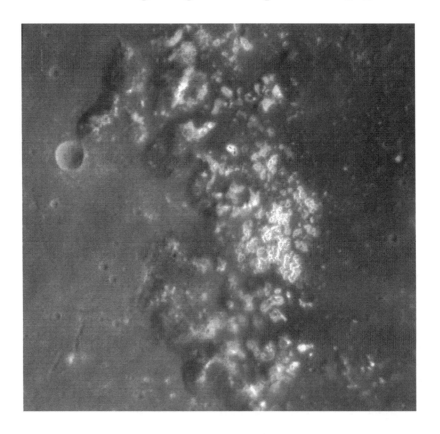

Figure 12. NASA MESSENGER image, taken with the Narrow Angle Camera, shows an area of hollows on the floor of Raditladi basin on Mercury.

Surface hollows were first discovered on Mercury during MESSENGER's orbital mission and have not been seen on the Moon or on any other rocky planetary body. This high-reflectance material, I posit, formed as hydrogen from Mercury's core exited the planet and reduced iron sulfide to the metal. Verifying that the bright metal is in fact low-nickel iron metal will provide strong evidence for high-pressure, high-temperature protoplanetary formation, and concomitant planetocentric nuclear reactor formation.

8.0 Perspectives

The chemical state and location of thorium in the deep-interior of the Earth is unknown. Like uranium, thorium occurs mainly in that portion of the Abee enstatite chondrite that corresponds to the Earth's core [45]. But very little thorium can exist within the nuclear sub-core; otherwise, the absorption of neutrons by thorium would poison the nuclear fission chain reaction. So, a fundamental question arises as to whether thorium resides in the georeactor nuclear waste sub-shell or elsewhere in the core. If thorium resides in the georeactor nuclear waste sub-shell, then the question becomes whether thorium interaction with neutrons will breed additional fuel, i.e., fissionable ^{233}U.

Detection of antineutrinos, also called geoneutrinos, which originate deep within the Earth has the potential for verifying georeactor existence and for revealing the location of uranium and thorium. Already geoneutrino detection has established upper limits on the georeactor nuclear fission contribution of either 26% [75] or 15% [74] of the total energy output of uranium and thorium. Improving detection efficiency and lowering background will be beneficial, but will not delineate actinide locations; directional detectors will be required, as well as verification that geoneutrino attenuation with depth is insignificant as presently believed.

The georeactor is not unique to the Earth. Its existence is the consequence of planetary formation by raining-out at high temperatures and pressures in the interior of a giant gaseous protoplanet. The commonality of that mode of planetary formation, yielding highly reduced condensed in the Solar System, and the commonality of micro-gravity georeactor operating environments makes understandable georeactor-type magnetic field generation in most planets and large moons [27, 50].

Many of the images from the MESSENGER spacecraft, like Figure 12, reveal '... an unusual landform on Mercury, characterized by irregular shaped, shallow, rimless depressions, commonly in clusters and in association with high-reflectance material ... and suggest that it indicates recent volatile-related activity' [91]. But the authors, reasoning within the framework of consensus-approved models, were unable to describe a scientific basis for the source of those volatiles or to suggest identification of the 'high-reflectance material'. By contract, I [92] calculated that, during condensation at pressures ≥ 1 atm, copious amounts of hydrogen, one or more Mercury volumes at STP could be incorporated in Mercury's fluid iron alloy core, which will be released as the core subsequently solidifies. Hydrogen geysers, exiting the surface, formed the pits or hollows and are possibly involved in the exhalation of iron sulfide, which is abundant on the planet's surface, and some of which may have been reduced to iron metal thus accounting for the associated 'high-reflectance material', bright spots. Verifying that the 'high-reflectance material' is indeed metallic iron of low nickel content will not only provide strong evidence for Mercury's hydrogen geysers, but more generally will provide evidence that planetary interiors 'rained out' by condensing within giant-gaseous protoplanets at high pressures, which is the same circumstance responsible for causing uranium to reside in the core and function as a planetocentric nuclear fission reactor.

The agreement between observed ^{3}He/^{4}He in deep-source oceanic basalt and the numerical simulation helium fission products is strong evidence for georeactor existence. The calculated progressive increase in the ratio over time (Figure 8) is particularly significant in light of ^{3}He/^{4}He > 10 R_A observed in hotspot 'mantle plumes' that have been seismically imaged to the edge of the core. It should be possible in principle to observe the change in ^{3}He/^{4}He over time by measurement along volcanic chains, such as the Hawaiian-Emperor volcanic chain, as was initially measured by Keller *et al.* [93]. Their results suggest a progressive increase in ^{3}He/^{4}He since the Cretaceous. Clearly, further work is needed for better statistics and to more clearly delineate mineral hosts that best preserve the helium signature over time.

Nuclear fission consumes uranium fuel; eventually, the georeactor fuel supply will become depleted, thermal convection will cease in nuclear waste sub-shell, and the geomagnetic field will collapse [24]. When will georeactor demise occur? That is currently unknown, but can in principle, and should be addressed through the measuring the temporal change in deep-source helium ratios. It is important to determine the time of georeactor demise, because of the low mass of the georeactor, one ten-millionth that of the core, the end may come quickly with potentially devastating effects on humanity.

Acknowledgements

I have benefitted from conversations with P. K. Iyengar, Paul K. Kuroda, Inge Lehmann, Lynn Margulis, Hans E. Suess, and Harold C. Urey. I thank Daniel F. Hollenbach and Oak Ridge National Laboratory personnel who generously devised a methodology to allow their software, originally designed for man-made nuclear reactors, to utilize geological time scales and to simulate fission product removal [28].

References

1. Hahn, O. and Strassmann, F., Uber den nachweis und das verhalten der bei der bestrahlung des urans mittels neutronen entstehenden erdalkalimetalle. *Die Naturwissenschaften,* 1939, **27**, 11-15.

2. Meitner, L. and Frisch, O.R., Disintegration of uranium by neutrons: A new type of nuclear reaction. *Nature,* 1939, **143**, 239-249.

3. Libby, W.F., Stability of uranium for natural fission. *Phys. Rev.,* 1939, **55**, 1236.

4. Flerov, G.N. and Petrzhak, K.A., Spontaneous fission of uranium. *Phys. Rev.,* 1940, **58**, 89.

5. Petrzhak, K.A. and Flerov, G.N., Über die spontane Teilung von Uran. *Compt. Rend. Akad. Sci. USSR,* 1940, **25**, 500.

6. Flügge, F., Kann der energieinhalt der atomkerne technisch nutzbar gemacht werden? *Die Naturwissenschaften,* 1939, **27**, 402.

7. Noetzlin, J., Volcanism and nuclear chemistry. *Comptes rendus,* 1939, **208**, 1662-1664.

8. Orr, J.B., Uranium in thucholite. *Phys. Rev.,* 1950, **76**, 155.

9. Fleming, W.H. and Thode, H.G., Neutron and spontaneous fission in uranium ores. *Phys. Rev.,* 1953, **92**, 378.

10. Wetherill, G.W., Spontaneous fission yields from uranium and thorium. *Phys. Rev.,* 1953, **92**, 907-912.

11. Wetherill, G.W. and Inghram, M.G., In *Nuclear Processes in Geological Settings*, eds., National Research Council and National Science Foundation, Williams Bay, Wisconsin.

12. Kuroda, P.K., On the nuclear physical stability of the uranium minerals. *J. Chem. Phys.,* 1956, **25**, 781-782.

13. Kuroda, P.K., On the infinite multiplication constant and the age of uranium minerals. *J. Chem. Phys.,* 1956, **25**, 1295-1296.

14. Fermi, E., Elementary theory of the chain-reacting pile. *Sci.,* 1947, **105**, 27-32.

15. Cowan, G.A., A natural fission reactor. *Sci. Am.,* 1976, **235**, 36-47.

16. Maurette, M., Fossil nuclear reactors. *A. Rev. Nuc. Sci.,* 1976, **26**, 319-350.

17. Meshik, A.P., Hohenberg, C.M. and Pravdivtseva, O.V., Record of cycling operation of the natural nuclear reactor in the Oklo/Okelobondo area in Gabon. *Phys. Rev. Lett,* 2004, **93**, 182302.

18. Hagemann, R. and Roth, E., Relevance of the studies of the Oklo natural reactors to the storage of radioactive wastes. *Radiochemica Acta,* 1978, **25**, 241-247.

19. Gauthier-Lafaye, F., Holliger, P. and Blanc, P.-L., Natural fission reactors in the Franceville basin, Gabon: A review of the conditions and results of a "critical event" in a geologic system. *Geochim. Cosmochim. Acta,* 1996, **60**, 4831-4852.

20. Herndon, J.M., Feasibility of a nuclear fission reactor at the center of the earth as the energy source for the geomagnetic field. *J. Geomag. Geoelectr.,* 1993, **45**, 423-437.

21. Herndon, J.M., Planetary and protostellar nuclear fission: Implications for planetary change, stellar ignition and dark matter. *Proc. R. Soc. Lond.,* 1994, **A455**, 453-461.

22. Herndon, J.M., Sub-structure of the inner core of the earth. *Proc. Nat. Acad. Sci. USA,* 1996, **93**, 646-648.

23. Herndon, J.M., Examining the overlooked implications of natural nuclear reactors. *Eos, Trans. Am. Geophys. U.,* 1998, **79**, 451,456.

24. Herndon, J.M., Nuclear georeactor origin of oceanic basalt ^{3}He/^{4}He, evidence, and implications. *Proc. Nat. Acad. Sci. USA,* 2003, **100**, 3047-3050.

25. Herndon, J.M., Solar system processes underlying planetary formation, geodynamics, and the georeactor. *Earth, Moon, and Planets,* 2006, **99**, 53-99.

26. Herndon, J.M., Nuclear georeactor generation of the earth's geomagnetic field. *Curr. Sci.,* 2007, **93**, 1485-1487.

27. Herndon, J.M., Nature of planetary matter and magnetic field generation in the solar system. *Curr. Sci.,* 2009, **96**, 1033-1039.

28. Hollenbach, D.F. and Herndon, J.M., Deep-earth reactor: Nuclear fission, helium, and the geomagnetic field. *Proc. Nat. Acad. Sci. USA,* 2001, **98**, 11085-11090.

29. Herndon, J.M., *Maverick's Earth and Universe.* Trafford Publishing, Vancouver, 2008.

30. Herndon, J.M., *Indivisible Earth: Consequences of earth's early formation as a jupiter-like gas giant.* Ed., Margulis, L., Thinker Media, Inc.

31. Herndon, J.M., *Origin of the geomagnetic field: Consequence of earth's early formation as a jupiter-like gas giant*. Thinker Media, Inc.

32. Wiechert, E., Über die Massenverteilung im inneren der Erde. *Nachr. K. Ges. Wiss. Göttingen, Math.-Kl.,* 1897, 221-243.

33. Oldham, R.D., The constitution of the interior of the earth as revealed by earthquakes. *Q. T. Geol. Soc. Lond.,* 1906, **62**, 456-476.

34. Lehmann, I., P'. *Publ. Int. Geod. Geophys. Union, Assoc. Seismol., Ser A, Trav. Sci.,* 1936, **14**, 87-115.

35. Scott, E.R.D., Chondrites and the protoplanetary disc. *Ann. Rev. Earth Planet. Sci.,* 2007, **35**, 577-620.

36. Birch, F., The transformation of iron at high pressures, and the problem of the earth's magnetism. *Am. J. Sci.,* 1940, **238**, 192-211.

37. Baedecker, P.A. and Wasson, J.T., Elemental fractionations among enstatite chondrites. *Geochim. Cosmochim. Acta,* 1975, **39**, 735-765.

38. Jarosewich, E., Chemical analyses of meteorites: A compilation of stony and iron meteorite analyses. *Meteoritics,* 1990, **25**, 323-337.

39. Kallcmcyn, G.W., et al., Ordinary chondrites: Bulk compositions, classification, lithophile-element fractionations, and composition-petrographic type relationships. *Geochim. Cosmochim. Acta,* 1989, **53**, 2747-2767.

40. Kallemeyn, G.W. and Wasson, J.T., The compositional composition of chondrites-I. The carbonaceous chondrite groups. *Geochim. Cosmochim. Acta,* 1981, **45**, 1217-1230.

41. Herndon, J.M., The nickel silicide inner core of the earth. *Proc. R. Soc. Lond.,* 1979, **A368**, 495-500.

42. Ramdohr, P., Einiges über Opakerze im Achondriten und Enstatitachondriten. *Abh. D. Akad. Wiss. Ber., Kl. Chem., Geol., Biol.,* 1964, **5**, 1-20.

43. Reed, S.J.B., Perryite in the Kota-Kota and South Oman enstatite chondrites. *Mineral Mag.,* 1968, **36**, 850-854.

44. Ringwood, A.E., Silicon in the metal of enstatite chondrites and some geochemical implications. *Geochim. Cosmochim. Acta,* 1961, **25**, 1-13.

45. Murrell, M.T. and Burnett, D.S., Actinide microdistributions in the enstatite meteorites. *Geochim. Cosmochim. Acta,* 1982, **46**, 2453-2460.

46. Dziewonski, A.M. and Anderson, D.A., Preliminary reference earth model. *Phys. Earth Planet. Inter.,* 1981, **25**, 297-356.

47. Keil, K., Mineralogical and chemical relationships among enstatite chondrites. *J. Geophys. Res.,* 1968, **73**, 6945-6976.

48. Kennet, B.L.N., Engdahl, E.R. and Buland, R., Constraints on seismic velocities in the earth from travel times. *Geophys. J. Int.,* 1995, **122**, 108-124.

49. Herndon, J.M., Geodynamic basis of heat transport within the earth *Curr Sci,* 2011, **101**, 1440-1450.

50. Herndon, J.M., New indivisible planetary science paradigm. *Curr Sci,* 2013, **105**, 450-460.

51. SCALE: A Modular Code System for Performing Standardized Analyses for Licensing Evaluations, N.C.-., Rev. 4, (ORNL/NUREG/CSD-2/R4), Vols. I, II, and III, April, 1995. Available from Radiation Safety Information Computational Center at Oak Ridge National Laboratory as CCC-545.

52. DeHart, M.D. and Hermann, O.W., An extension of the validation of SCALE (SAS2H) isotopic predictions for PWR spent fuel,

ORNL/TM-13317, Lockheed Martin Energy Research Corp., Oak Ridge National Laboratory. 1996.

53. England, T.R., Wilson, R.E., Schenter, R.E. and Mann, F.M., Summary of ENDF/B-V data for fission products and actinides, EPRI NP-3787 (LA-UR 83-1285) (ENDF-322), Electric Power Research Institute.

54. Hermann, O.W., San onofre pwr data for code validation of MOX fuel depletion analyses, ORNA/TM-1999/018, R1, Lockheed Martin Energy Research Corp., Oak Ridge National Laboratory. 2000.

55. Hermann, O.W., Bowman, S.M., Brady, M.C. and Parks, C.V., Validation of the SCALE system for PWR spent fuel isotopic composition analyses, ORNL/TM-12667, Martin Marietta Energy Systems, Oak Ridge National Laboratory. 1995.

56. Hermann, O.W. and DeHart, M.D., Validation of SCALE (SAS2H) isotopic predictions for BWR spent fuel, ORNL/TM-13315, Lockheed Martin Energy Research Corp., Oak Ridge National Laboratory. 1998.

57. Seifritz, W., Some comments on Herndon's nuclear georeactor. *Kerntechnik,* 2003, **68**, 193-196.

58. Herndon, J.M. and Edgerley, D.A., Background for terrestrial antineutrino investigations: Radionuclidc distribution, gcorcactor fission events, and boundary conditions on fission power production. *arXiv:hep-ph/0501216 24 Jan 2005,* 2005.

59. de Meijer, R.L. and van Westrenen, W., The feasibility and implications of nuclear georeactors in earth's core–mantle boundary region. *S. Af. J. Sci.,* 2008, **104**, 111-118.

60. Anisichkin, V.F., Bezborodov, A.A. and Suslov, I.R., Georeactor in the earth. *Transport. Theo. Stat. Phys.,* 2008, **37**, 624-633.

61. Rusov, V.D., Pavlovich, V.N., Vaschenko, V.N., Tarasov, V.A., Zalentsova, T.N., Bolshakov, V.N., Litvinov, D.A., Kosenko, S.I. and

Byegunova, O.A., Geoantineutrino spectrum and slow nuclear burning on the boundary of the liquid and solid phases of the earth's core. *J. Geophys. Res.*, 2007, **112**.

62. Rao, K.R., Nuclear reactor at the core of the earth! - a solution to the riddles of relative abundances of helium isotopes and geomagnetic field variability. *Curr. Sci.*, 2002, **82**, 126-127.

63. Clarke, W.B., Beg, M.A. and Craig, H., Excess he-3 in the sea: Evidence for terrestrial primordial helium. *Earth Planet. Sci. Lett.*, 1969, **6**, 213-220.

64. Hilton, D.R., Grönvold, K., Macpherson, C.G. and Castillo, P.R., Extreme He-3/He-4 ratios in northwest iceland: Constraining the common component in mantle plumes. *Earth Planet. Sci. Lett.*, 1999, **173**, 53-60.

65. Anderson, D.L., The statistics of helium isotopes along the global spreading ridge system and the central limit theorem. *Geophys. Res. Lett.*, 2000, **27**, 2401-2404.

66. Bijwaard, H. and Spakman, W., Tomographic evidence for a narrow whole mantle plume below iceland. *Earth Planet. Sci. Lett.*, 1999, **166**, 121-126.

67. Nataf, H.-C., Seismic imaging of mantle plumes. *Ann. Rev. Earth Planet. Sci.*, 2000, **28**, 391-417.

68. Mjelde, R. and Faleide, J.I., Variation of icelandic and hawaiian magmatism: Evidence for co-pulsation of mantle plumes? *Mar. Geophys. Res.*, 2009, **30**, 61-72.

69. Mjelde, R., Wessel, P. and Müller, D., Global pulsations of intraplate magmatism through the cenozoic. *Lithosphere*, 2010, **2**, 361-376.

70. Cowan, J., C. L., Reines, F., Harrison, F.B., Krause, H.W. and McGuire, A.D., Detection of free neutrinos: A confirmation. *Sci.*, 1956, **124**, 103.

71. Raghavan, R.S. et al., Measuring the global radioactivity in the earth by multidectector antineutrino spectroscopy. *Phys. Rev. Lett.,* 1998, **80**, 635-638.

72. Raghavan, R.S., Detecting a nuclear fission reactor at the center of the earth. *arXiv:hep-ex/0208038,* 2002.

73. Domogatski, G., Kopeikin, L., Mikaelyan, L. and Sinev, V., Neutrino geophysics at baksan i: Possible detection of georeactor antineutrinos. *arXiv:hep-ph/0401221 v1* 2004.

74. Bellini, G. et al., Observation of geo-neutrinos. *Phys. Lett.,* 2010, **B687**, 299-304.

75. Gando, A. et al., Partial radiogenic heat model for earth revealed by geoneutrino measurements. *Nature Geosci.,* 2011, **4**, 647-651.

76. Pollack, H.N., Hurter, S.J. and Johnson, J.R., Heat flow from the earth's interior: Analysis of the global data set. *Rev. Geophys.,* 1993, **31**, 267-280.

77. McDonough, W.F. and Sun, S.-S., The composition of the Earth. *Chem. Geol.,* 1995. **120**, 223-253.

78. Herndon, J.M., A new basis of geoscience: Whole-earth decompression dynamics. *New Concepts in Global Tectonics Journal,* 2013, **1**, 81-95.

79. Herndon, J.M., Whole-earth decompression dynamics. *Curr. Sci.,* 2005, **89**, 1937-1941.

80. Herndon, J. M., Energy for geodynamics: Mantle decompression thermal tsunami. *Curr. Sci.,* 2006, **90**, 1605-1606.

81. Herndon, J. M., Beyond Plate Tectonics: Consequence of Earth's Early Formation as a Jupiter-Like Gas Giant, 2012, Thinker Media, Inc.

82. Herndon, J. M., Origin of mountains and primary initiation of submarine canyons: the consequences of Earth's early formation as a Jupiter-like gas giant. *Curr. Sci.*, 2012. **102**(10), 1370-1372.

83. Basu, A.R., Poreda, R.J., Renne, P.R., Teichmann, F., Vasiliev, Y.R., Sobolev, N.V. and Turrin, B.D., High-[3]He plume origin and temporal-spacial evolution of the Siberian flood basalts. *Sci.*, 1995, **269**, 882-825.

84. Courtillot, V., *Evolutionary* Catastrophes,2002, Cambridge University Press.

85. Herndon, J.M., Impact of recent discoveries on petroleum and natural gas exploration: Emphasis on India. *Curr. Sci.*, 2010, **98**, 772-779.

86. Elsasser, W.M., On the origin of the earth's magnetic field. *Phys. Rev.*, 1939, **55**, 489-498.

87. Elsasser, W.M., Induction effects in terrestrial magnetism. *Phys. Rev.*, 1946, **69**, 106-116.

88. Elsasser, W.M., The earth's interior and geomagnetism. *Revs. Mod. Phys.*, 1950, **22**, 1-35.

89. Bogue, S.W., Very rapid geomagnetic field change recorded by the partial remagnetization of a lava flow *Geophys. Res. Lett.*, 2010, **37**, doi: 10.1029/2010GL044286.

90. Coe, R.S. and Prevot, M., Evidence suggesting extremely rapid field variation during a geomagnetic reversal. *Earth Planet. Sci. Lett.*, 1989, **92**, 192-198.

91. Blewett, D. T., et al., Hollows on Mercury: MESSENGER Evidence for Geologically Recent Volatile-Related Activity. *Sci.*, 2011, **333**, 1859-1859.

92. Herndon, J. M., Hydrogen geysers: Explanation for observed evidence of geologically recent volatile-related activity on Mercury's surface. *Curr. Sci.*, 2012. **103**(4), 361-361.

93. Keller, R.A., et al., Cretaceous-to-recent record of elevated 3He/4He along the Hawaiian-Emperor volcanic chain. *Geochemistry, Geophysics, Geosystems*, 2004, **5**(12), 1-10.

Appendix C

New Indivisible Planetary Science Paradigm

J. Marvin Herndon
Transdyne Corporation

Published in

Current Science, Vol. 105(4), 25 August 2013, pp. 450-460.
Download PDF http://www.nuclearplanet.com/0450.pdf

Abstract: A new, indivisible planetary science paradigm, a wholly self-consistent vision of the nature of matter in the Solar System, and dynamics and energy sources of planets is presented here. Massive-core planets formed by condensing and raining-out from within giant gaseous protoplanets at high pressures and high temperatures. Earth's complete condensation included a ~300 Earth-mass gigantic gas/ice shell that compressed the rocky kernel to about 66% of Earth's present diameter. T-Tauri eruptions stripped the gases away from the inner planets and stripped a portion of Mercury's incompletely condensed protoplanet and transported it to the region between Mars and Jupiter where it fused with in-falling oxidized condensate from the outer regions of the Solar System and formed the parent matter of ordinary chondrite meteorites, the main-Belt asteroids and veneer for the inner planets, especially Mars. In response to decompression-driven planetary volume increases, cracks form to increase surface area and mountain ranges characterized by folding form to accommodate changes in curvature. The differences between the inner planets are primarily the consequence of different degrees of protoplanetary compression. The internal composition of Mercury is calculated by analogy with Earth. The rationale is provided for Mars potentially having a greater subsurface water reservoir capacity than previously realized.

Introduction

Images and data from orbiting spacecraft and landers have revealed new, important, unanticipated aspects of planets other than Earth. Understanding those observations, however, has posed a challenge for planetary investigators who generally are self-constrained within the framework of 'consensus favoured' models they consider applicable to the formation of the terrestrial planets, in particular, the so-called 'standard model of solar system formation' dating from the 1960s, and a model of the internal composition of Earth which had its beginning circa 1940.

Interpretations of other planets are strongly coloured by interpretations of our own, better-studied planet. In 1936, Lehmann discovered Earth's inner core [1]. At the time there was widespread belief that Earth resembled an ordinary chondrite meteorite. Within that circa 1940 understanding, the inner core's composition was thought to be iron metal in the process of crystallizing from the fluid iron alloy core [2]; the geomagnetic field was thought to be generated by convection-driven dynamo action in the fluid core, and; the rocky mantle surrounding the core was assumed to be of uniform composition with observed seismic discontinuities assumed to be caused by pressure-induced changes in crystal structure. Planetary investigators apply this interpretation of Earth to other planets, such as Mercury, but it is an incorrect interpretation.

I realized that discoveries made in the 1960s admitted a different possibility for the composition of the inner core, namely, fully crystallized nickel silicide [3]. That insight led me: (1) to evidence that Earth resembles, not an ordinary chondrite, but an enstatite chondrite; (2) to a fundamentally different interpretation of the composition of Earth's internal shells below a depth of 660 km and their state of oxidation; (3) to evidence of a new, powerful energy source and a different proposal for the generation-location of the geomagnetic field, and; (4) to a different understanding of Earth's formation and to new geodynamics that is the consequence. I have described the details and implications of this new, indivisible geoscience paradigm, called Whole-Earth Decompression Dynamics (WEDD), in a number of scientific articles [4-13] and books [14-18]. 'Indivisible' in this instance means that the fundamental aspects of Earth are connected logically and causally, and can be deduced from our planet's early formation as a Jupiter-like gas giant.

The visionary evolutionist, Lynn Margulis, taught the importance of envisioning the Earth as a whole, rather than as unrelated segments spread among various scientific specialties [15]. In that spirit, and in the broader framework of the Solar System, I present here a new, indivisible planetary science paradigm, a wholly self-consistent vision of the nature of matter in the Solar System, and dynamics and energy sources of planets [5-8, 12, 13, 19-21], which differs profoundly from the half-century old, popular, but problematic paradigm. This is a new foundation from which much development is possible.

Problematic Planetary Science Paradigm

The first hypothesis about the origin of the Sun and the planets was advanced in the latter half of the 18th Century by Immanuel Kant and modified later by Pierre-Simon de Laplace. Early in the 20th Century, Laplace's nebula hypothesis was replaced with the Chamberlin-Moulton hypothesis which held that a passing star pulled matter from the Sun which condensed into large protoplanets and small planetesimals. Although the passing star idea fell out of favour, the nomenclature of protoplanets and planetesimals remained. Generally, concepts of planetary formation fall into one of two categories that involve either (1) condensation at high-pressures, hundreds to thousands of atmospheres (atm.) or (2) condensation at very low pressures.

Eucken [22] considered the thermodynamics of Earth condensing and raining-out within a giant gaseous protoplanet at pressures of 100-1000 atm. In the 1950s and early 1960s there was discussion of planetary formation at such pressures [23-25], but that largely changed with the 1963 publication by Cameron [26] of a model of solar system formation from a primordial gas of solar composition at low pressure, circa 10^{-4} atm.. Cameron's low pressure model became the basis for (1) condensation models that (wrongly) purported to produce minerals characteristic of ordinary chondrites as the equilibrium condensate from that medium [27, 28] and (2) planetary formation models based upon the Chamberlin-Moulton planetesimal hypothesis. The idea was that dust would condense from the gas at this very low pressure. Dust grains would collide with other grains, sticking together to become progressively larger grains, then pebbles, then rocks, then planetesimals and finally planets [29, 30].

Since the 1960s, the planetary science community almost unanimously concurred that Earth formed from primordial matter that condensed at a very low pressure, circa 10^{-4} atm. [27, 31]. The 'planetesimal hypothesis' was 'accepted' as the 'standard model of solar system formation'. However, as I discovered, there is an inherent flaw in that concept [5, 8, 32].

All the inner planets have massive cores, as known from their high relative densities. I was able to show by thermodynamic calculations that the condensate of primordial matter at those very low pressures would be oxidized, like the Orgueil C1/CI meteorite wherein virtually all elements are combined with oxygen. In such low pressure, low temperature condensate, there would be essentially no iron metal for the massive cores of the inner planets, a contradiction to the observation of massive-core planets.

The planetesimal hypothesis, *i.e.*, the 'standard model of solar system formation', is not only problematic from the standpoint of planetary bulk-density, but necessitates additional ad hoc hypotheses. One such necessary hypothesis is that of a radial Solar System temperature gradient during planetary formation, an assumed warm inner region delineated by a hypothetical 'frost line' between Mars and Jupiter; ice/gas condensation is assumed to occur only beyond that frost line. Another such necessary hypothesis is that of whole-planet melting, i.e., the 'magma ocean', to account for core formation from essentially undifferentiated material. For other planetary systems with close-to-star gas giants, another such necessary hypothesis is that of 'planetary migration' where gas giants are assumed to form at Jupiter-distances from their star and then migrate inward.

Primary Mode of Planetary Formation

The popular version of planetary formation described above consists of an assemblage of assumption-based hypotheses that lack substantive connection with one another. That is not the case in the new, indivisible planetary science paradigm presented here: The highly-reduced state of primitive enstatite-chondrite matter is explained by high-pressure, high-temperature condensation from solar matter [5, 33] under circumstances similar to those derived by Eucken [22] for Earth raining out from within a giant gaseous protoplanet and the relative masses of inner parts of Earth, derived from seismic data, match corresponding, chemically-identified, relative masses of enstatite-chondrite-components (Table 1, Figure 1), observed by microscopic examination, indicating commonality of oxidation state and formation process.

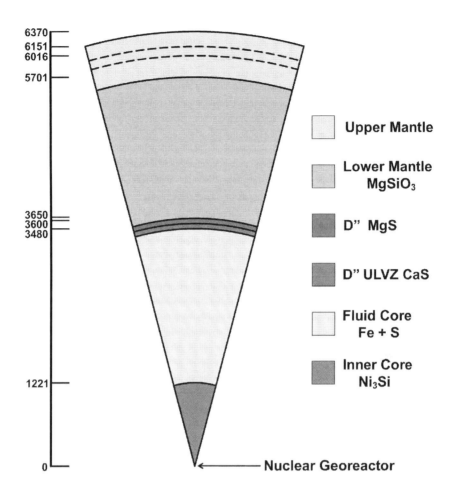

Figure 1. Chemical compositions of the major parts of the Earth, inferred from the Abee enstatite chondrite (see Table 1). The upper mantle, above the lower mantle, has seismically-resolved layers whose chemical compositions are not yet known. Radial distance scale in km.

Table 1. Fundamental mass ratio comparison between the endo-Earth (lower mantle plus core) and the Abee enstatite chondrite. Above a depth of 660 km seismic data indicate layers suggestive of veneer, possibly formed by the late addition of more oxidized chondrite and cometary matter, whose compositions cannot be specified with certainty at this time.

Fundamental Earth Ratio	Earth Ratio Value	Abee Ratio Value
lower mantle mass to total core mass	1.49	1.43
inner core mass to total core mass	0.052	theoretical 0.052 if Ni_3Si 0.057 if Ni_2Si
inner core mass to lower mantle + total core mass	0.021	0.021
D'' mass to total core mass	0.09[***]	0.11[*]
ULVZ[**] of D'' CaS mass to total core mass	0.012[****]	0.012[*]

[x] = avg. of Abee, Indarch, and Adhi-Kot enstatite chondrites
D'' is the "seismically rough" region between the fluid core and lower mantle
[**] ULVZ is the "Ultra Low Velocity Zone" of D''
[***] calculated assuming average thickness of 200 km
[****] calculated assuming average thickness of 28 km
data from [55-57]

Thermodynamic considerations led Eucken [22] to conceive of Earth formation from within a giant, gaseous protoplanet when molten iron rained out to form the core, followed by the condensation of the silicate-rock mantle. By similar, extended calculations I verified Eucken's results and deduced that oxygen-starved, highly-reduced matter characteristic of enstatite chondrites and by inference the Earth's interior, condensed at high temperatures and high pressures from primordial Solar System gas under circumstances that isolated the condensate from further reaction with the gas at low temperatures [5, 33].

In primordial matter of solar composition, there is a relationship between condensation pressure, condensation temperature, and the state of oxidation of the condensate. Ideally, when the partial pressure of a particular substance in the gas exceeds the vapor pressure of that condensed substance, the substance will begin to condense. In a gas of solar composition, the partial pressure of a substance is directly proportional to the total gas pressure, so at higher pressures substances condense at higher temperatures. The degree of oxidation of the condensate, on the other hand, is determined by the gas phase reaction

$$\textbf{H}_2 + \textbf{½O}_2 \leftrightarrow \textbf{H}_2\textbf{O}$$

which is a function of temperature but essentially independent of pressure. As I discovered, that reaction leads to an oxidized condensate at low temperatures and to a highly-reduced condensate at high temperatures, provided the condensate is isolated from further reaction with the gas [5, 33].

At pressures above about 1 atm. in a primordial atmosphere of solar composition, iron metal condenses as a liquid (Figure 2). That liquid can dissolve and sequester certain other elements, including significant hydrogen and a portion of oxygen-loving elements such as Ca, Mg, Si, and U. The composition and structure of the Earth's core (Figure 1) can be understood from the metallurgical behaviour of an iron alloy of this composition initially with all of the core-elements fully dissolved at some high-temperature.

Elements with a high affinity for oxygen are generally incompatible in an iron alloy. So, when thermodynamically feasible those elements escaped from the liquid alloy. Calcium and magnesium formed CaS and MgS, respectively, which floated to the top of the core and formed the region referred to as D''. Silicon combined with nickel, presumably as Ni_3Si, and formed the inner core. The trace element uranium precipitated, presumably as US, and through one or more steps settled at the center of the Earth where it engaged in self-sustaining nuclear fission chain reactions [5, 20, 34-37].

The gaseous portion of primordial Solar System matter, as is the Sun's photosphere today, was about 300 times as massive as all of its rock-plus-metal forming elements. I posited Earth's complete condensation formed a gas-giant planet virtually identical in mass to Jupiter [4, 8, 15]. Giant gaseous planets of Jupiter size are observed in other planetary systems as close or closer to their star than Earth is to the Sun [38].

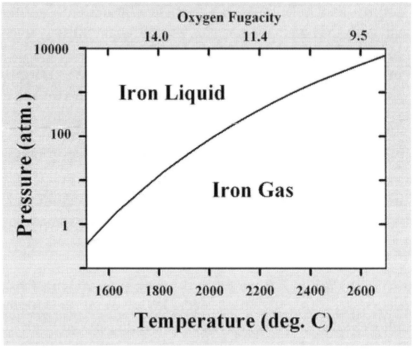

Figure 2. The curve in this figure shows the temperatures and total pressures in a cooling atmosphere of solar composition at which liquid iron will ideally begin to condense. The pressure-independent oxygen fugacity is shown on the upper abscissa.

Of the eight planets in the Solar System, the outer four (Jupiter, Saturn, Uranus, and Neptune) are gas-giants, whereas the inner four are rocky (Mercury, Venus, Earth, and Mars), without primary atmospheres. However, the inner planets originated from giant gaseous protoplanets and their massive, primordial gases. How were the gases lost?

A brief period of violent activity, the T-Tauri phase, occurs during the early stages of star formation with grand eruptions and super-intense "solar-wind". The Hubble Space Telescope image of an erupting binary T-Tauri star is shown in Figure 3. The white crescent shows the leading edge of the plume from a five-year earlier observation. The plume edge moved 130AU, a distance 130 times that from the Sun to Earth, in just 5 years. A T-Tauri outburst by our young Sun, I posit, stripped gas from the inner four planets. A rocky Earth, compressed by the weight of primordial gases, remained. Eventually, Earth began to decompress driven primarily by the stored energy of protoplanetary compression. The consequences of Earth's formation in this manner provide rich new ways to interpret planetary data, especially when viewed in the broader context of Solar System processes responsible for the diversity of planet-forming matter.

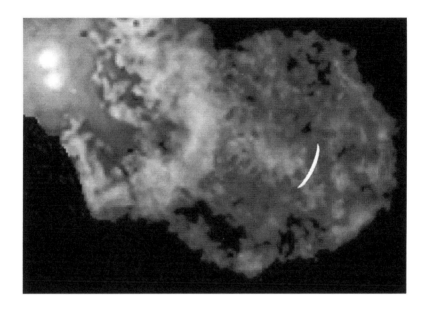

Figure 3. Hubble Space Telescope image of binary star XZ-Tauri in 2000 showing a T-Tauri phase outburst. The white crescent label shows the position of the leading edge of that plume in 1995, indicating a leading-edge advance of 130 A.U. in five years. T-Tauri eruptions are observed in newly formed stars. Such eruptions from our nearly-formed Sun, I submit, stripped the primordial gases from the inner four planets of our Solar System.

Matter of the Asteroid Belt, Mercury, and Ordinary Chondrites

The near-constancy in isotopic compositions of most of the elements of the Earth, the Moon, and the meteorites indicates formation from primordial matter of common origin [32]. Exceptions do occur and are important cosmochemical tracers, for example, oxygen and, in refractory inclusions of carbonaceous chondrites, magnesium, silicon, calcium, and titanium. Primordial elemental composition is yet evident to a great extent in the photosphere of the Sun and, for the less volatile, rock-forming elements, in chondrite meteorites, where many elements have not been separated from one another to within a factor of two. However, there is complexity: rather than just one type of chondrite, there are three, with each type characterized by its own strikingly unique state of oxidation. Understanding the nature of the processes that yielded those three distinct types of matter from one common progenitor forms the basis for understanding much about planetary formation, their compositions, and the processes they manifest, including magnetic field production.

Only five major elements, iron (Fe), magnesium (Mg), silicon (Si), oxygen (O), and sulfur (S), comprise at least 95% of the mass of each chondrite and, by implication, each of the terrestrial planets. For decades, the abundances of major rock-forming elements (E_i) in chondrites have been expressed in the literature as atom ratios, usually relative to silicon (E_i/Si) and occasionally relative to magnesium (E_i/Mg). By expressing major-element abundances as molar (atom) ratios relative to iron (E_i/Fe), I discovered a fundamental relationship bearing on the genesis of chondrite matter, shown in Figure 4, which has implications on the nature of planetary processes in the Solar System [7]. Note in Figure 4 that the ordinary chondrite line intersects the other two. For this unique circumstance, each ordinary chondrite can be expressed as a linear combination of the compositions at the points of intersection. One intersection-component is a relatively undifferentiated carbonaceous-chondrite-like primitive component, with a state of oxidation like the Orgueil C1/CI chondrite, while the other is a partially differentiated enstatite-chondrite-like planetary component.

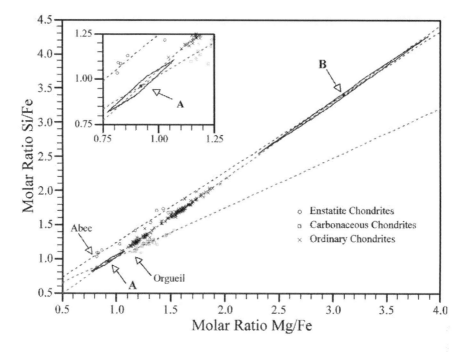

Figure 4. Molar (atom) ratios of Mg/Fe and Si/Fe from analytical data on 10 enstatite chondrites, 39 carbonaceous chondrites, and 157 ordinary chondrites. Data from [58-60]. Members of each chondrite class data set scatter about a unique, linear regression line. Upper line, enstatite chondrites; lower line carbonaceous chondrites, and; intersecting line, ordinary chondrites. The locations of the volatile-rich Orgueil carbonaceous chondrite and the volatile-rich Abee enstatite chondrite are indicated. Line intersections A and B are designated, respectively, *primitive* and *planetary* components. Error estimates of points A and B are indicated by solid-line parallelograms formed from the intersections of the standard errors of the respective linear regression lines. Inset shows in expanded detail the standard error parallelogram of point A.

Ordinary chondrites possess the common characteristic of being markedly depleted in refractory siderophile elements such as iridium and osmium. The degree of iridium and osmium depletion in each ordinary chondrite correlates with the relative proportion of its planetary component [7]. One can therefore conclude that the planetary component originated from a single large reservoir, characterized by a depletion in iridium and in osmium. From the inferred composition of the planetary component indicated in Figure 4, I suggested the partially differentiated planetary component might be comprised of matter stripped from the protoplanet of incompletely-formed Mercury, presumably by the T-Tauri outbursts during thermonuclear ignition of the Sun. In the region between Mars and Jupiter, the ejected Mercury-component fused with in-falling Orgueil-like matter that had condensed at low pressures and low temperatures in the far reaches of the Solar System and/or in interstellar space. That fused combination become the parent matter of ordinary chondrites and asteroids of that region.

The molar (atom) $Mg/Fe = 3.1$ deduced for the planetary component indicates the stripping of Mercury's protoplanetary gases took place during the time when Mercury was only partially formed. The idea of heterogeneous protoplanetary differentiation/accretion is not new. Eucken [22] first suggested Earth's core formation as a consequence of successive condensation on the basis of relative volatility from a hot, gaseous protoplanet, with iron metal raining out at the center. The approximately seven-fold greater depletion within the planetary component of refractory siderophile elements (iridium and osmium) than other more volatile siderophile elements (nickel, cobalt and gold) indicates that planetary-scale differentiation and/or accretion progressed in a heterogeneous manner. The first liquid iron to condense and rain-out preferentially scavenged the refractory siderophile elements from the hot gaseous protoplanet.

I estimated the original total mass of ordinary chondrite matter present in the Solar System as a function of the core mass of Mercury [7]. For a core mass equal to 75% of Mercury's present mass, the calculated original total ordinary chondrite mass amounts to 1.83×10^{24} kg, about 5.5 times the mass of Mercury. That amount of mass is insufficient to have formed a planet as massive as the Earth, but may have contributed significantly to the formation of Mars, as well as adding a veneer to other planets, including Earth. Presently, only about 0.1% of that mass remains in the asteroid belt.

During the formation of the Solar System only three processes were primarily responsible for the diversity of matter in the Solar System and were directly responsible for planetary internal compositions and structures [5]. These are: (i) High-pressure, high-temperature condensation from primordial matter associated with planetary formation by raining-out from the interiors of giant-gaseous protoplanets; (ii) Low pressure, low temperature condensation from primordial matter in the remote reaches of the Solar System and/or in the interstellar medium associated with comets and (iii) Stripping of the primordial volatile components from the inner portion of the Solar System by super-intense T-Tauri phase outbursts during the thermonuclear ignition of the Sun. The internal composition of massive-core planets derives from (i) above, and leads to a simple commonality of highly-reduced internal planetary compositions. The outer portions of the terrestrial planets, however, appear in varying degree to be 'painted' by an additional veneer of more-oxidized matter derived from (ii) and (iii) above.

Inner Planets: Basis of Differences

Earth's surface is markedly different from that of the other inner planets in two pronounced ways: (1) About 41% of Earth's surface area is comprised of continental rock (sial) with the balance being ocean floor basalt (sima), and (2) Like stitching on a baseball, a series of mid-ocean ridges encircles the Earth from which basalt extrudes, creeps across the ocean basins, and disappears into trenches. As disclosed in Whole-Earth Decompression Dynamics (WEDD), these are consequences of Earth's early formation as a Jupiter-like gas giant with the rocky portion initially compressed to about 66% of present diameter by about 300 Earth-masses of primordial gases and ices [4].

Surface differences among the inner planets, I posit, are the consequence of circumstances that prevented the rocky kernels of other inner planets from being fully compressed by condensed gigantic gas/ice shells. As described above, stripping of Mercury's protoplanetary gases is inferred to have taken place during the time when Mercury was only partially formed [7]. One might speculate from relative density that the rocky kernel of Venus was fully formed, but the extent of its compression may differ from that of Earth due to the prevailing thermal environment and/or relative time of the Sun's T-Tauri outbursts. Eventually, the degree of compression experienced should be able to be estimated by understanding Venetian surface geology. Mars may be a special circumstance, having a relatively small, highly-reduced kernel surrounded by a relatively large shell of ordinary chondrite matter; additional information is needed to be more precise.

Earth's crust is markedly different from that of the other inner planets in harbouring a geothermal gradient. Similar to Earth's two-component crust, the otherwise inexplicable geothermal gradient is understandable as a consequence of our planet's early formation as a Jupiter-like gas giant.

Since 1939, scientists have been measuring the heat flowing out of continental-rock [39, 40] and, since 1952, heat flowing out of ocean floor basalt [41]. Continental-rock contains much more of the long-lived radioactive nuclides than does ocean floor basalt. So, when the first heat flow measurements were reported on continental-rock, the heat was assumed to arise from radioactive decay. However, later, ocean floor heat flow measurements, determined far from mid-ocean ridges, showed more heat flowing out of the ocean floor basalt than out of continental-rock measured away from heat-producing areas [42, 43]. This seemingly paradoxical result, I posit, arises from a previously unanticipated mode of heat transport that emplaces heat at the base of the crust. I call this mode of heat transport Mantle Decompression Thermal Tsunami [6].

Heat generated deep within the Earth may enhance mantle decompression by replacing the lost heat of protoplanetary compression. The resulting decompression, beginning within the mantle, will tend to propagate throughout the mantle, similar to a tsunami, until it reaches the impediment posed by the base of the crust. There, crustal rigidity opposes continued decompression; pressure builds and compresses matter at the mantle-crust-interface resulting in compression heating. This compression heating, I submit, is the source of heat that produces the geothermal gradient.

Earth's geothermal gradient serves as a barrier that limits the downward migration of water. The "geothermal gradient" is minimal or non-existent for terrestrial planets that lack the compression-stage characterized by an early, massive, fully condensed shell of primordial gases and ices. Mars appears to have lacked an early massive shell of compressive condensed gases. Without subsequent decompression of the Martian kernel, there is no basis to assume the existence of a "geothermal gradient"; there is no thermal barrier to the downward percolation of water. The absence of such a thermal barrier suggests that Mars may have a much greater subsurface water reservoir potential than previously realized.

In the popular, problematic planetary science paradigm, internal planetary heat is produced through the decay of long-lived radionuclides, the only non-hypothetical heat source, although for moons sometimes tidal friction is also included. In the new, indivisible planetary science paradigm described here, the following two important energy sources are added: (1) Stored energy of protoplanetary compression which, in the case of Earth, is the principle driving-energy for decompression and for heat emplacement at the base of the crust by *Mantle* Decompression Thermal Tsunami, and; (2) Planetocentric 'georeactor' nuclear fission energy.

During Earth's early formation as a Jupiter-like gas giant, the weight of ~300 Earth-masses of gas and ice compressed the rocky kernel to approximately 66% of present diameter. Owing to rheology and crustal rigidity, the protoplanetary energy of compression was locked-in when the T-

Tauri outbursts stripped away the massive gas/ice layer leaving behind a compressed kernel whose crust consisted entirely of continental rock (sial). Internal pressures began to build and eventually the crust began to crack.

To accommodate decompression-driven increases in volume in planetary volume, Earth's surface responds in two fundamentally different ways; by crack formation and by the formation of mountain chains characterized by folding.

Cracks form to increase the surface area required as a consequence of planetary-volume increases. Primary cracks are underlain by heat sources and are capable of basalt extrusion, for example, mid-ocean ridges; secondary cracks are those without heat sources, for example, submarine trenches, and which become the ultimate repositories for basalt extruded by primary cracks.

In addition to crack formation, decompression-increased planetary volume necessitates adjustments in surface curvature. Decompression-driven increases in volume result in a misfit of the continental rock surface formed earlier at a smaller Earth-diameter. This misfit results in 'excess' surface material confined within continent margins, which adjusts to the

Figure 5. Demonstration illustrating the formation of fold-mountains as a consequence of Earth's early formation as a Jupiter-like gas giant. On the left, two balls representing the relative proportions of 'present' Earth (large), and 'ancient' Earth (small) before decompression. In the center, a spherical section, representing a continent, cut from 'ancient' Earth and placed on the 'present' Earth, showing: (1) the curvature of the 'ancient continent' does not match the curvature of the 'present' Earth and (2) the 'ancient continent' has 'extra' surface area confined within its fixed perimeter. On the right, tucks remove 'extra' surface area and illustrate the process of fold-mountain formation that is necessary for the 'ancient' continent to conform to the curvature of the 'present' Earth. Unlike the ball-material, rock is brittle so tucks in the Earth's crust would break and fall over upon themselves producing fold-mountains.

new surface curvature by buckling, breaking and falling over upon itself producing fold-mountain chains as illustrated in Figure 5 from ref. [12].

Crack formation and the production of mountains characterized by folding, consequences of protoplanetary compression, are pronounced processes on Earth and may have some relevance to Venus. Planetocentric 'georeactor' nuclear fission energy, on the other hand, has relevance to virtually all planets and to some large moons.

Evidence from Mercury's Surface

One of the most important Project MESSENGER discoveries were images from the spacecraft that revealed '... an unusual landform on Mercury, characterized by irregular shaped, shallow, rimless depressions, commonly in clusters and in association with high-reflectance material ... and suggest that it indicates recent volatile-related activity' (Figure 6) and which have not been observed on any other rocky planet [44]. However, planetary investigators were unable to describe a scientific basis for the source of those volatiles or to suggest identification of the 'high-reflectance material'. I posited that during formation, condensing and raining-out as a liquid at high pressures and high temperatures from within a giant gaseous protoplanet, Mercury's iron alloy core dissolved copious amounts of hydrogen, one or more Mercury-volumes at STP. Hydrogen is quite soluble in liquid iron, but much less soluble in solid iron. I suggested that dissolved hydrogen from Mercury's core, released during core-solidification and escaping at the surface, produced hydrogen geysers that were responsible for forming those 'unusual landform on Mercury', sometimes referred to as pits or hollows, and for forming the associated 'high-reflectance material', bright spots, which I suggested is iron metal reduced from an exhaled iron compound, probably iron sulfide, by the escaping hydrogen [13].

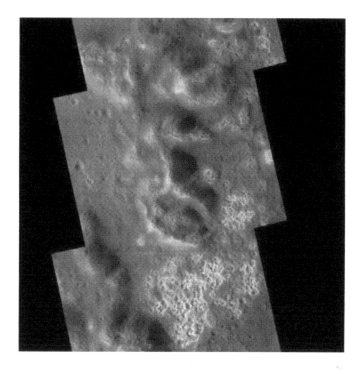

Figure 6. NASA MESSENGER image, taken with the Narrow Angle Camera, shows an area of hollows on the floor of Raditladi basin on Mercury.

Surface hollows were first discovered on Mercury during MESSENGER's orbital mission and have not been seen on the Moon or on any other rocky planetary bodies. These bright, shallow depressions appear to have been formed by disgorged volatile material(s) from within the planet.

So, here is a test: Verifying that the 'high-reflectance material' is indeed metallic iron will not only provide strong evidence for Mercury's hydrogen geysers, but more generally will provide evidence that planetary interiors rained-out by condensing at high pressures and high temperatures within giant gaseous protoplanets. The high reflectance metallic iron can be distinguished by its low-nickel content from meteoritic metallic iron.

By analogy with Earth (Figure 1), the compositions of the interior parts of Mercury, calculated according to the mass ratio relationships presented in Table 1, are shown in Figure 7. Mercury's $MgSiO_3$ mantle mass is taken as the difference between planet mass and calculated core mass. Only nine elements account for about 98% of the mass of a chondrite meteorite and the planet Mercury. Of the major and minor elements comprising Mercury's core, depicted in Figure 7, only aluminum and sodium, which have a high affinity for oxygen, are not represented. Presumably all aluminum and most, if not all, sodium occurs in Mercury's mantle/crust. Possibly a minor amount sodium might occur in Mercury's core as $NaCrS_2$ [45]. In the extreme case, if all of the trace element Cr formed $NaCrS_2$, a maximum of 18% of Mercury's sodium might occur as $NaCrS_2$.

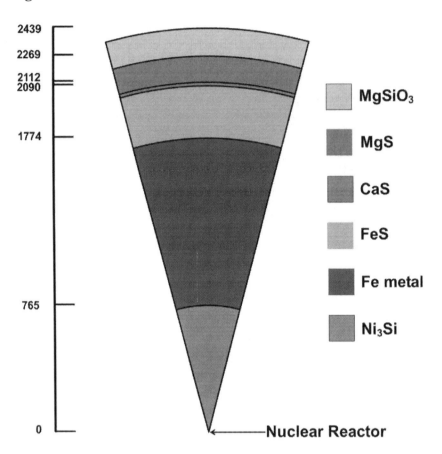

Figure 7. Internal structure of Mercury calculated from the mass ratio relationships of Earth shown in Table 1. Mercury's core is assumed to be fully solidified. The initial location of the planetocentric 'georeactor' is indicated. Radial distance scale in km.

As with Earth, the composition and structure of the Mercury's core (Figure 7) can be understood from the metallurgical behavior of an iron alloy initially with all of the core-elements fully dissolved at some high temperature. Upon cooling sufficiently, calcium and magnesium formed CaS and MgS, respectively, which floated to the top of the Mercurian core and formed the region analogous to Earth's D''. Silicon combined with nickel, presumably as Ni_3Si, and formed the inner core. The trace element uranium precipitated, presumably as US, and through one or more steps settled at

Mercury's centre where it inevitably engaged in self-sustaining nuclear fission chain reactions [8].

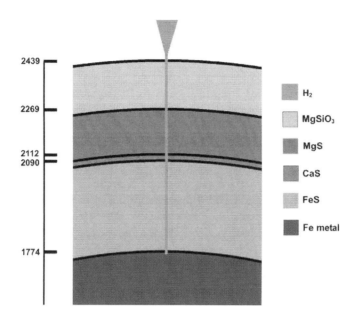

Figure 8. Schematic illustration of the source and path of hydrogen which is exhausted as hydrogen geysers and forms hollows (pits) on Mercury's surface. Radial distance scale in km.

One of the surprising early discoveries of the Project MESSENGER mission was abundant sulfur on Mercury's surface [46]. That observation is understandable as a consequence of hydrogen geysers. Figure 8 is a schematic representation of the path taken by exsolved hydrogen. Note the exiting hydrogen gas traverses regions of various sulfide compositions: iron sulfide (FeS), calcium sulfide (CaS) and magnesium sulfide (MgS). The exiting hydrogen, I submit, may scavenge sulfides from these layers and deposit them on Mercury's surface and perhaps may even emplace some in Mercury's exosphere.

Mercury is about 6% as massive as Earth. In the 1970s, this tiny planet's core, based upon heat-flow calculations, was thought to have solidified within the first billion years after formation [47]. However, that was before my demonstration of the feasibility of planetocentric nuclear fission reactors [8, 21, 34] whose energy production considerably delayed solidification. Later, upon subsequent cooling, iron metal began to precipitate from Mercury's iron-sulfur alloy fluid core; the endpoint of core solidification is depicted in Figure 7. Core solidification with its concomitant release of dissolved hydrogen provides explanations for Mercurian surface phenomena.

Commonality of Nuclear Fission Heat and Magnetic Field Generation

Internally generated, currently active magnetic fields have been detected in six planets (Mercury, Earth, Jupiter, Saturn, Uranus and Neptune) and in one satellite (Jupiter's moon Ganymede). Magnetized surface areas of Mars and the Moon indicate the former existence of internally generated magnetic fields in those bodies. Furthermore, Jupiter, Saturn and Neptune radiate about twice as much energy as each receives from the Sun. Energy from nuclear fission chain reactions, part of the new, indivisible planetary science paradigm described here, provides logical and causally related explanations [8].

The condensate from within a giant gaseous protoplanet resembles an enstatite chondrite; thermodynamic condensation considerations are similar [5, 22, 33]. The interior of Earth, below 660 km, resembles an enstatite chondrite (Table 1). Thus, one may reasonably conclude that the Earth formed by raining out from within a giant gaseous protoplanet and that the interiors of other planets are similar to Earth's interior, which means their interiors are highly-reduced like the Abee enstatite chondrite. In the Abee meteorite, uranium occurs in the non-oxide part that corresponds to the Earth's core.

In cores of planets, density is a function of atomic number and atomic mass. Uranium, being the densest substance would tend ultimately to accumulate at the planets' centre. Applying Fermi's nuclear reactor theory, I demonstrated the feasibility of planetocentric nuclear fission reactors as energy sources for Jupiter, Saturn and Neptune [19, 20] and for Earth as the energy source for the geomagnetic field [20, 34, 35]. Numerical simulations subsequently made at Oak Ridge National Laboratory verified those calculations and demonstrated that the georeactor could function over the entire age of the Earth as a fast neutron breeder reactor [36, 37]. Moreover, the calculations showed that helium would be produced in precisely the range of isotopic compositions observed exiting Earth.

The georeactor is a two-part assemblage, as illustrated in Figure 9, consisting of a fissioning nuclear sub-core surrounded by a sub-shell of radioactive waste products, presumably a liquid or slurry. The ~24 km diameter assemblage is too small to be presently resolved from seismic data.

Oceanic basalt helium data, however, provide strong evidence for the georeactor's existence [36, 48] and antineutrino measurements have not refuted that [49, 50]. To date, detectors at Kamioka, Japan and at Gran Sasso, Italy have detected antineutrinos coming from within the Earth. After years of data-taking, an upper limit on the georeactor nuclear fission contribution was determined to be either 26% (Kamioka, Japan) [50] or 15% (Gran Sasso, Italy) [49] of the total energy output of uranium and thorium, estimated from deep-Earth antineutrino measurements (Table 2). The actual total georeactor contribution may be somewhat greater, though, as some georeactor energy comes from natural decay as well as from nuclear fission.

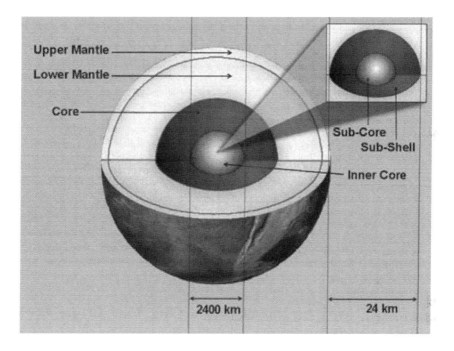

Figure 9. Earth's nuclear fission georeactor (inset) shown in relation to the major parts of Earth.

The georeactor at the center is one ten-millionth the mass of Earth's fluid core. The georeactor sub-shell, I posit, is a liquid or a slurry and is situated between the nuclear-fission heat source and inner-core heat sink, assuring stable convection, necessary for sustained geomagnetic field production by convection-driven dynamo action in the georeactor sub-shell [8, 21, 35].

Before the Mariner 10 flybys in 1974 and 1975, in light of predictions of early core-solidification [47], there was essentially no expectation that Mercury possesses a currently generated magnetic field. That changed. The

MESSENGER observations confirmed the existence of an actively generated, albeit very weak, global magnetic field centered close to the spin axis [51]. Efforts to explain Mercury's magnetic field generation within the problematic planetary science paradigm have proven to be challenging. This is why: Popular cosmochemical models fashioned on the idea that the internal composition of Mercury resembles an ordinary chondrite do not predict a substantial source of heat in Mercury's core. Without such a heat source, the core would solidify within about one billion years thus rendering core-convection impossible [47].

Table 2. Geoneutrino (antineutrino) determinations of radiogenic heat production [49, 50] shown for comparison with Earth's heat loss to space [61]. See original report for discussion and error estimates.

Heat (terawatts)	Source
44.2 TW	global heat loss to space
20.0 TW	neutrino contribution from ^{238}U, ^{232}Th, and georeactor fission
5.2 TW	georeactor KamLAND data
3.0 TW	georeactor Borexino data
4.0 TW	^{40}K theoretical
20.2 TW	loss to space minus radiogenic

In 1939, Elsasser first published his idea that the geomagnetic field is produced by convective motions in the Earth's fluid, electrically conducting core, interacting with rotation-produced Coriolis forces, creating a dynamo mechanism, a magnetic amplifier [52-54]. Elsasser's convection-driven dynamo mechanism seemed to explain so well the generation of the geomagnetic field that for decades geophysicists believed convection in the Earth's fluid core 'must' exist. Later, when it was discovered that many planets had internally generated magnetic fields, they were assumed, by analogy to Earth, to have convecting fluid iron alloy cores. However, there is a problem, not with Elsasser's idea of a convection-driven dynamo, but with its location; as I discovered, convection is physically impossible in the Earth's fluid core and, presumably, as well in the cores of the various planets [8, 11].

I have suggested the convection-driven dynamo produces the geomagnetic field in the georeactor sub-shell, which has implications for magnetic field reversals. The mass of the georeactor is only one ten-millionth the mass of the fluid core. High-intensity changing outbursts of solar wind, through the intermediary of the geomagnetic field, will induce electric currents into the georeactor, causing ohmic heating, which in extreme cases, might disrupt dynamo-convection and lead to a magnetic reversal. Massive trauma to the Earth might also disrupt sub-shell convection and lead to a magnetic reversal.

Why is there no Earth-core convection? The core is bottom-heavy, being approximately 23% denser at the bottom than at the top. The small decrease in density at the bottom due to thermal expansion is insufficient to overcome such a great density gradient. Moreover, for sustained convection the core-top must be maintained at a lower temperature than the core-bottom which is impossible because the Earth's core is wrapped in the mantle, a 2900 km thick thermally insulating blanket that has considerably lower thermal conductivity and heat capacity than the core.

In the popular problematic planetary science paradigm another problem is evident: There is no basis for the existence of a central heat source to drive the assumed planetary-core convection. However, in the new, indivisible planetary science paradigm described here, all of those problems are moot.

I have suggested that the geomagnetic field is produced by Elsasser's convection-driven dynamo operating within the georeactor's radioactive waste sub-shell [21]. Unlike the Earth's core, sustained convection appears quite feasible in the georeactor sub-shell. The top of the georeactor sub-shell is in contact with the inner core, a massive heat sink, which is in contact with the fluid core, another massive heat sink. Heat brought from the nuclear sub-core to the top of the georeactor sub-shell by convection is efficiently removed by these massive heat sinks thus maintaining the sub-shell adverse temperature gradient. Moreover, the sub-shell is not bottom heavy. Further, decay of neutron-rich radioactive waste in the sub-shell provides electrons that might provide the seed magnetic fields for amplification.

Among massive-core planets and large moons, there is a commonality of formation by condensing and raining-out of a gas of solar composition at high temperatures and high pressures, which leads to a commonality of internal compositions and highly-reduced states of oxidation, which in turn leads to a commonality of georeactor-like planetocentric nuclear fission reactors. In each case, the central nuclear reactor is about one ten-millionth as massive as the planet's core and its operation does not depend upon the physical state of the core. The small mass means that major impacts could in principle offset the nuclear core from the planets centre which, for example, might explain why Mercury's magnetic field is offset ~484 km north of center [51].

Venus currently has no internally generated magnetic field. Four potential explanations are: (1) Venus' rotation rate may be too slow; (2) Venus currently may be experiencing interrupted sub-shell convection such as might occur during a magnetic reversal; (3) Fuel breeding reactions at some point may have been insufficient for continued reactor operation, or (4) Venus' 'georeactor' may have consumed all of its fissionable fuel. In light of helium evidence portending the eventual demise of Earth's georeactor [36], the fourth explanation seems most reasonable.

Summary

Massive-core planets formed by condensing and raining-out from within giant gaseous protoplanets at high pressures and high temperatures, accumulating heterogeneously on the basis of volatility with liquid core-formation preceding mantle-formation; the interior states of oxidation resemble that of the Abee enstatite chondrite. Core-composition was established during condensation based upon the relative solubilities of elements, including uranium, in liquid iron in equilibrium with an atmosphere of solar composition at high pressures and high temperatures. Uranium settled to the central region and formed planetary nuclear fission reactors, producing heat and planetary magnetic fields.

Earth's complete condensation included a ~300 Earth-mass gigantic gas/ice shell that compressed the rocky kernel to about 66% of Earth's present diameter. T-Tauri eruptions, associated with the thermonuclear ignition of the Sun, stripped the gases away from the Earth and the inner planets. The T-Tauri outbursts stripped a portion of Mercury's incompletely condensed protoplanet and transported it to the region between Mars and Jupiter where it fused with in-falling oxidized condensate from the outer regions of the Solar System and/or interstellar space, forming the parent matter of ordinary chondrite meteorites, the main-Belt asteroids, and veneer for the inner planets, especially Mars.

With its massive gas/ice shell removed, pressure began to build in the compressed rocky kernel of Earth and eventually the rigid crust began to crack. The major energy source for planetary decompression and for heat emplacement at the base of the crust is stored energy of protoplanetary compression. In response to decompression-driven volume increases, cracks form to increase surface area and fold-mountain ranges form to accommodate changes in curvature.

One of the most profound mysteries of modern planetary science is this: As the terrestrial planets are more-or-less of common chondritic composition, how does one account for the marked differences in their surface dynamics? Differences among the inner planets are principally due to the degree of compression experienced. Planetocentric georeactor nuclear fission, responsible for magnetic field generation and concomitant heat production, is applicable to compressed and non-compressed planets and large moons.

The internal composition of Mercury is calculated based upon an analogy with the deep-Earth mass ratio relationships. The origin and implication of Mercurian hydrogen geysers is described. Besides Earth, only Venus appears to have sustained protoplanetary compression; the degree of which might eventually be estimated from understanding Venusian surface geology. A basis is provided for understanding that Mars essentially lacks a 'geothermal gradient' which implies potentially greater subsurface water reservoir capacity than previously expected.

References

1. Lehmann, I., P'. *Publ. Int. Geod. Geophys. Union, Assoc. Seismol., Ser. A, Trav. Sci.*, 1936, **14**, 87-115.

2. Birch, F., The transformation of iron at high pressures, and the problem of the earth's magnetism. *Am. J. Sci.*, 1940, **238**, 192-211.

3. Herndon, J. M., The nickel silicide inner core of the Earth. *Proc. R. Soc. Lond*, 1979, **A368**, 495-500.

4. Herndon, J. M., Whole-Earth decompression dynamics. *Curr. Sci.*, 2005, **89**(10), 1937-1941.

5. Herndon, J. M., Solar System processes underlying planetary formation, geodynamics, and the georeactor. *Earth, Moon, and Planets*, 2006, **99**(1), 53-99.

6. Herndon, J. M., Energy for geodynamics: Mantle decompression thermal tsunami. *Curr. Sci.*, 2006, **90**, 1605-1606.

7. Herndon, J. M., Discovery of fundamental mass ratio relationships of whole-rock chondritic major elements: Implications on ordinary chondrite formation and on planet Mercury's composition. *Curr. Sci.*, 2007, **93**(3), 394-398.

8. Herndon, J. M., Nature of planetary matter and magnetic field generation in the solar system. *Curr. Sci.*, 2009, **96**, 1033-1039.

9. Herndon, J. M., Impact of recent discoveries on petroleum and natural gas exploration: Emphasis on India. *Curr. Sci.*, 2010, **98**, 772-779.

10. Herndon, J. M., Potentially significant source of error in magnetic paleolatitude determinations. *Curr. Sci.*, 2011, **101**(3), 277-278.

11. Herndon, J. M., Geodynamic Basis of Heat Transport in the Earth. *Curr. Sci.*, 2011, **101**, 1440-1450.

12. Herndon, J. M., Origin of mountains and primary initiation of submarine canyons: the consequences of Earth's early formation as a Jupiter-like gas giant. *Curr. Sci.*, 2012. **102**(10), 1370-1372.

13. Herndon, J. M., Hydrogen geysers: Explanation for observed evidence of geologically recent volatile-related activity on Mercury's surface. *Curr. Sci.*, 2012. **103**(4), 361-361.

14. Herndon, J. M., *Maverick's Earth and Universe,* 2008, Vancouver: Trafford Publishing. ISBN 978-1-4251-4132-5.

15. Herndon, J. M., *Indivisible Earth: Consequences of Earth's Early Formation as a Jupiter-Like Gas Giant,* L. Margulis, Editor 2012, Thinker Media, Inc.

16. Herndon, J. M., Beyond Plate Tectonics: Consequence of Earth's Early Formation as a Jupiter-Like Gas Giant, 2012, Thinker Media, Inc.

17. Herndon, J. M., *Origin of the Geomagnetic Field: Consequence of Earth's Early Formation as a Jupiter-Like Gas Giant,* 2012, Thinker Media, Inc.

18. Herndon, J. M., *What Meteorites Tell Us About Earth,* 2012, Thinker Media, Inc.

19. Herndon, J. M., Nuclear fission reactors as energy sources for the giant outer planets. *Naturwissenschaften*, 1992, **79**. 7-14.

20. Herndon, J. M., Planetary and protostellar nuclear fission: Implications for planetary change, stellar ignition and dark matter. *Proc. R. Soc. Lond*, 1994, **A455**, 453-461.

21. Herndon, J. M., Nuclear georeactor generation of the earth's geomagnetic field. *Curr. Sci.*, 2007, **93**(11), 1485-1487.

22. Eucken, A., Physikalisch-chemische Betrachtungen ueber die frueheste Entwicklungsgeschichte der Erde. *Nachr. Akad. Wiss. Goettingen, Math.-Kl.*, 1944, 1-25.

23. Bainbridge, J., Gas imperfections and physical conditions in gaseous spheres of lunar mass. *Astrophys. J.*, 1962, **136**, 202-210.

24. Kuiper, G. P., On the origin of the Solar System. *Proc. Nat. Acad. Sci. USA*, 1951, **37**, 1-14.

25. Kuiper, G. P., On the evolution of the protoplanets. *Proc. Nat. Acad. Sci. USA*, 1951, **37**, 383-393.

26. Cameron, A. G. W., Formation of the solar nebula. *Icarus*, 1963, **1**, 339-342.

27. Grossman, L., Condensation in the primitive solar nebula. *Geochim. Cosmochim. Acta*, 1972, **36**, 597-619.

28. Larimer, J. W. and Anders, E., Chemical fractionations in meteorites-III. Major element fractionations in chondrites. *Geochim. Cosmochim. Acta*, 1970, **34**, 367-387.

29. Goldrich, P. and Ward, W. R., The formation of planetesimals. *Astrophys J.*, 1973, **183**(3),1051-1061.

30. Wetherill, G. W., Formation of the terrestrial planets. *Ann. Rev. Astron. Astrophys.*, 1980, **18**: p. 77-113.

31. Larimer, J. W., Chemistry of the solar nebula. *Space Sci. Rev.*, 1973, **15**(1), 103-119.

32. Herndon, J. M., Reevaporation of condensed matter during the formation of the solar system. *Proc. R. Soc. Lond*, 1978, **A363**, 283-288.

33. Herndon, J. M. and Suess, H. E., Can enstatite meteorites form from a nebula of solar composition? *Geochim. Cosmochim. Acta*, 1976, **40**, 395-399.

34. Herndon, J. M., Feasibility of a nuclear fission reactor at the center of the Earth as the energy source for the geomagnetic field. *J. Geomag. Geoelectr.*, 1993, **45**, 423-437.

35. Herndon, J. M., Sub-structure of the inner core of the earth. *Proc. Nat. Acad. Sci. USA*, 1996, **93**, 646-648.

36. Herndon, J. M., Nuclear georeactor origin of oceanic basalt ^{3}He/^{4}He, evidence, and implications. *Proc. Nat. Acad. Sci. USA*, 2003, **100**(6), 3047-3050.

37. Hollenbach, D. F. and Herndon, J. M., Deep-earth reactor: nuclear fission, helium, and the geomagnetic field. *Proc. Nat. Acad. Sci. USA*, 2001, **98**(20), 11085-11090.

38. Seager, S. and Deming, D., Exoplanet Atmospheres. *Ann. Rev. Astron. Astrophys.*, 2010, **48**, 631-672.

39. Benfield, A. F., Terrestrial heat flow in Great Britain. *Proc. R. Soc. Lond*, 1939, **A173**, 428-450.

40. Bullard, E. C., Heat flow in South Africa. *Proc. R. Soc. Lond*, 1939, **A173**, 474-502.

41. Revelle, R. and Maxwell, A. E., Heat flow through the floor of the eastern North Pacific Ocean. Nature, 1952, **170**, 199-200.

42. Blackwell, D. D., The thermal structure of continental crust, in *The Structure and Physical Properties of the Earth's Crust, Geophysical Monograph 14*, J. G. Heacock, Editor 1971, American Geophysical Union: Washington, DC., p. 169-184.

43. Stein, C. and Stein, S., A model for the global variation in oceanic depth and heat flow with lithospheric age. *Nature*, 1992, **359**, 123-129.

44. Blewett, D. T., et al., Hollows on Mercury: MESSENGER Evidence for Geologically Recent Volatile-Related Activity. *Sci.*, 2011, **333**, 1859-1859.

45. Okada, A. and Keil, K., Caswellsilverite, NaCrS$_2$: A new mineral in the Norton County enstatite achondrite. *Am. Min.*, 1982, **67**, 132-136.

46. Nittler, L. R., et al., The major element composition of Mercury's surface from MESSENGER X-ray spectrometry. *Sci.*, 2011, **333**, 1847-1850.

47. Solomon, S. C., Some aspects of core formation in Mercury. *Icarus*, 1976, **28**(4), 509-521.

48. Rao, K. R., Nuclear reactor at the core of the Earth! - A solution to the riddles of relative abundances of helium isotopes and geomagnetic field variability. *Curr. Sci.*, 2002, **82**(2), 126-127.

49. Bellini, G., et al., Observation of geo-neutrinos. *Phys. Lett.*, 2010, **B687**, 299-304.

50. Gando, A., et al., Partial radiogenic heat model for Earth revealed by geoneutrino measurements. *Nature Geosci.*, 2011, **4**, 647-651.

51. Anderson, B. J., et al., The global magnetic field of Mercury from MESSENGER orbital observations. *Sci.*, 2011., **333**, 1859-1862.

52. Elsasser, W. M., On the origin of the Earth's magnetic field. *Phys. Rev.*, 1939. **55**, 489-498.

53. Elsasser, W. M., Induction effects in terrestrial magnetism. Phys. Rev., 1946, **69**, 106-116.

54. Elsasser, W. M., The Earth's interior and geomagnetism. *Revs. Mod. Phys.*, 1950, **22**, 1-35.

55. Dziewonski, A. M. and Anderson, D. A., Preliminary reference Earth model. *Phys. Earth Planet. Inter.*, 1981, **25**, 297-356.

56. Keil, K., Mineralogical and chemical relationships among enstatite chondrites. *J. Geophys. Res.*, 1968, **73**(22), 6945-6976.

57. Kennet, B. L. N., Engdahl, E. R. and Buland, R., Constraints on seismic velocities in the earth from travel times. *Geophys. J. Int.*, 1995, **122**, 108-124.

58. Baedecker, P.A. and Wasson, J. T., Elemental fractionations among enstatite chondrites. *Geochim. Cosmochim. Acta*, 1975, **39**, 735-765.

59. Jarosewich, E., Chemical analyses of meteorites: A compilation of stony and iron meteorite analyses. *Meteoritics*, 1990, **25**, 323-337.

60. Wiik, H. B., On regular discontinuities in the composition of meteorites. *Commentationes Physico-Mathematicae*, 1969, **34**, 135-145.

61. Pollack, H. N., Hurter, S. J., Johnson, J. R., Heat flow from the Earth's interior: Analysis of the global data set. *Rev. Geophys.*, 1993, **31**(3), 267-280.

Appendix D

Critique of Quality Bulletin for Peer Review

J. Marvin Herndon

Submitted to White House December 16, 2004

Introduction: From time to time a government might conduct a comprehensive review of a particular process and/or procedure and the report becomes the new "gold standard" for implementation of the particular process and/or procedure. But can self-appraisals really be objective? How often does a government really seriously question the very processes and procedures it employs? Even when public comment is solicited, generally only those insiders with vested interests make statements. Moreover, public comment is generally restricted to pre-report opinions that may or may not be given much weighed during government review. After the report is published, in my experience, the new "gold standard" is "cast in concrete"; comment and criticism is neither invited nor acted upon. But after publication, I submit, is the time for independent critique and recommendations.

On December 16, 2004, an individual in the White House to whom I had complained about the inequity of "peer review" sent me a copy of the Office of Management and Budget's "Final Information Quality Bulletin for Peer Review: December 15, 2004".

On December 26, 2004, I sent to the White House my critique of that Bulletin and my recommendations for systemic changes, which were neither appreciated nor implemented. Ten years later, the U. S. Government still conducts peer review according to that Bulletin, and American science continues to decline. Have responsible officials never heard the phrase: "Find out what's wrong and fix it"?

Critique of Bulletin: This Bulletin appears to have been crafted by individuals who either are extremely naïve of human nature or choose to ignore human nature. The Bulletin appears to be predicated upon the tacit assumption that in all instances peer reviewers will provide honest, truthful reviews. The tacit assumption that peer reviewers will always be truthful leads to a principal flaw of this Bulletin, namely, the failure to provide any instruction, direction, or requirement either to guard against fraudulent peer review or to prosecute those suspected of making untruthful reviews. The long-standing failure of the federal government to require and to aggressively enforce truthfulness in the making of peer reviews (1) encourages and rewards those who make deceitful peer reviews and (2) can be expected to have and to have had a deleterious impact on America's scientific capability, adversely affecting America's technology, and, concomitantly, weakening America's economic and military capability.

The Bulletin states, "... the names of each reviewer may be publicly disclosed or remain anonymous (*e.g.*, to encourage candor)." The Bulletin approves the application of anonymity and even appears to promote some alleged virtue of its use, while being completely blind to its downside. If anonymity did in fact encourage candor and truthfulness, anonymity would be of great value in our legal system. There is in fact a historical record of the use of anonymity in courts: Anonymous testimony was used in a historical record of the use of anonymity in courts: Anonymous testimony was used in the Spanish Inquisition and in nearly every totalitarian regime. In each instance, the results were the same: individuals denounce others, for a variety of reasons. Anonymity, instead of being a positive element, as implied by the Bulletin, is instead an extremely negative element, encouraging and rewarding the worst aspects of human nature and human behavior.

In the selection of participants in peer review, the Bulletin urges "due consideration of independence and conflict of interest" but, at the same time: (1) the Bulletin provides a highly restricted financial definition of conflict of interest, ignoring personal, professional, or scientific conflicts of interest, (2) the Bulletin falsely accords some individuals a conflict-of-interest-free-status, specifically, "...when a scientist is awarded a government research grant through an investigator-initiated, peer-reviewed competition, there generally should be no question as to that scientist's ability to offer independent scientific advice to the agency on other projects", (3) the Bulletin acknowledges that "the federal government recognize that under certain circumstances some conflict may be unavoidable..." and, (4) the Bulletin completely prevents the avoidance of conflicts of interest by approving the use of anonymous reviews, a practice which is wide-spread in federal government agencies that support scientific research.

The Bulletin fails to provide any instruction, direction, or requirement either to guard against fraudulent peer review or to prosecute those suspected of making untruthful reviews. At the same time, the Bulletin approves of the use of anonymous reviews that are free from accountability and from civil recourse. The Bulletin gives tacit approval to circumstances that allow conflicts of interest and prevents the avoidance of conflicts of interest. With that combination, the Bulletin encourages and gives free rein to any criminal or quasi-criminal element that seeks to attain unfair advantage through deceit and misrepresentation in the anonymity-protected peer review system.

The Bulletin states that "reviewers may be compensated for their work or they may donate their time as a contribution to science or public service". All instances with which I am familiar, specifically, NASA and the National Science Foundation, do not pay reviewers for their time. Reviewers do reviews for a variety of reasons, such as to curry favor with agency officials or to exercise control over their competitors, but the bottom line is that time is money and agencies often get for their non-compensation a hastily done, superficial review.

The most serious short-coming of the Bulletin is its failure to recognize or to admit the debilitating consequences of the long-term application of the practices described above. The application of anonymity and freedom from accountability in the peer review system and the openness to conflicts of interest, gives unfair advantage to those who would unjustly berate a competitor's formal request for research funding. The perception – real or imagined – that some individuals would do just that has had a chilling effect, forcing scientists to become defensive, adopting only the consensus-approved viewpoint and refraining from discussing anything that might be considered as a challenge to other's work or to the funding agency's programs. That is not science!

If a foreign power or a terrorist group had set out to slowly and imperceptibly undermine American science, I doubt that it could have devised a methodology for the purpose any more effective than the practices set forth in the Bulletin. Used for decades, these practices have diminished American science to the present point of approaching third-world status.

Recommendations for Systemic Changes in the Administration of Peer Review: Like the Critique, the following recommendations are specifically limited to the administration of peer review as applied to reviews of scientific grant/contract proposals that are not subject to secrecy considerations related to national security and defense.

> •Generally, conflicts of interest, in the broadest definition, should never be permitted and should never be tolerated in peer review. Federal regulations permitting same, such as those under which NASA operates, should be changed.

•Anonymity should never be used in peer reviews. Reviewers should sign and swear their reviews "under penalty of perjury" and should be held accountable for the truthfulness of their reviews.

•Reviewers should always be compensated, and compensated well, for making reviews.

•An independent "ombudsman agency" should be created to address conflicts and disagreements between the individual or organization submitting the research proposal and the agency to which the proposal was submitted. The "ombudsman agency" should be empowered to bring potentially unlawful activity to the attention of the Department of Justice for investigation and possible prosecution.

•Recipients of federal research grants/contracts should incur the responsibility of making some pre-determined number of reviews under the conditions described above. Like defense attorneys selecting a jury, each grants/contract recipient should be accorded a number of "pre-emptive strikes" so as to be able to remove himself/herself from certain specific proposal reviews.

•An individual or organization submitting a research proposal should be presented with an official list of names that the funding agency proposes to be peer reviewers. Like prosecuting attorneys selecting a jury, the individual or organization submitting the proposal should be accorded a number of "pre-emptive strikes" so as to be able to remove certain specific proposed reviewers.

Implementation of the above recommendations will begin to correct the long-standing debilitation of American science. The above recommendations are intended to correct the short-comings displayed in the Bulletin. Additional managerial improvements are possible, but are not specified in the present document.

Appendix E

Planetary and Protostellar Nuclear Fission: Implications for Planetary Change, Stellar Ignition and Dark Matter

J. Marvin Herndon
Herndon Science & Software, Inc.

Published in *Proceedings of the Royal Society of London*, A445, 453-461 (1994)

Abstract: The feasibility of thermal neutron fission and fast neutron fission in planetary and protostellar matter is calculated from nuclear reactor theory. Means for concentrating actinide elements and for separating actinide elements from reactor poisons are described. The implications of intermittent or interrupted planetaryscale nuclear fission breeder reactors are discussed in connection with observed changes in the giant outer planets and changes in the geomagnetic field. The concept that thermonuclear fusion reactions in stars are ignited by nuclear fission energy is disclosed. The suggestion is made that dark matter, inferred to exist in the Universe, might be accounted for, at least in part, by the presence of dark stars (not necessarily brown dwarfs) whose protostellar nuclear fission reactors failed to ignite thermonuclear fusion reactions.

1. Introduction

In 1939, Hahn & Strassmann (1939) reported their discovery of nuclear fission. Later in the same year, Flügge (1939) considered the possibility that self-sustaining chain reactions might have taken place under natural conditions sometime in the past within uranium ore deposits. Kuroda (1956) subsequently applied nuclear reactor theory (Fermi 1947) to demonstrate the feasibility that uranium ore deposits in nature might in the geological past have become critical and functioned as thermal neutron nuclear fission reactors. In 1972, French scientists discovered at Oklo in the Republic of Gabon, Africa, the fossil remains of an actual natural reactor (Baudin *et al.* 1972; Bodu *et al.* 1972; Neuilly *et al.* 1972). Recently, I developed the concept of planetary-scale natural nuclear fission reactors (Herndon 1992, 1993). This paper addresses the role of nuclear fission in the planetary and astronomical sciences.

2. Background

The giant planets, Jupiter, Saturn, and Neptune, radiate into space approximately twice as much energy as they receive from the Sun; Uranus, on the other hand, emits little energy other than absorbed solar energy (Pearl *et al.* 1990). Planetary scientists believed that they had considered all possible energy sources and concluded 'by elimination' that the excess emitted energy must be a relic left over from planetary formation about 4600 million years ago (Hubbard 1990). This view was first challenged by Herndon (1992) who suggested nuclear fission energy.

I have presented evidence for the occurrence of substantial quantities of uranium (and thorium) in the Earth's core and have demonstrated the feasibility for nuclear fission as an energy source for the geomagnetic field (Herndon 1993). Furthermore, I have suggested that polarity reversals of the geomagnetic field may have their origins in intermittent nuclear reactor output (Herndon 1993).

To my knowledge, there exists no observational data on protostars before the ignition of thermonuclear fusion reactions. However, the planet Jupiter represents in certain respects a reasonable protostar model (Hubbard 1990). Since before the discovery of nuclear fission, gravitational potential energy, released during protostellar collapse, has been assumed as the energy source for the ignition of thermonuclear fusion reactions in the stars (Bethe 1939; Gamow & Teller 1938; Leve 1953; Schwartzschild 1958). Protostar heating by the gravitational infall of matter is off-set by radiation from the surface which is a function of the fourth power of temperature. Generally, in numerical models of protostellar collapse, ignition temperatures, on the order of several million degrees Celsius, are not attained solely by the gravitational infall of matter; an additional shock wave induced sudden flare up is assumed (Hayashi & Nakano 1965; Larson 1984). The concept of planetary nuclear fission reactors, as applied to the giant gaseous planets and to the Earth's core (Herndon 1992, 1993), may also apply to protostars and forms the basis of the suggestion, made in this paper, that thermonuclear fusion reactions in stars, as in hydrogen bombs, are ignited by self-sustaining, neutron induced, nuclear fission.

3. Theoretical basis

The pressures that prevail in the deep interiors of planets are sufficiently great that the density of matter is essentially a function of atomic number and atomic mass (Herndon 1992). Actinide elements, being the most dense substances, would tend, by the action of gravity, to be concentrated at the planets' or protostars' centre and separated from less dense reactor poisons as shown by figures 1 and 2.

Figure 1 shows theoretical estimates of the density of several substances as a function of pressure calculated using a Thomas-Fermi-Dirac approach published by Salpeter & Zapolsky (1967). The pressure-density profile of a solar mixture of hydrogen and helium, applicable for example to Jupiter and to protostellar internal regions, designated $H_{10}He$ in figure 1, is one boundary-value reference. The pressure-density profile of nickeliferous iron, $Fe_{16}Ni$, is applicable to planetary cores, although the addition of lighter elements would certainly lead to a slight decrease in density. Nevertheless, the $Fe_{16}Ni$ curve serves as a useful reference for comparing the pressure-density profiles of actinides, represented in figure 1 by uranium mono-sulphide and uranium metal. Fission-product reactor poisons, as represented by the example of ^{149}Sm in figure 1, are less dense than uranium or compounds of uranium at all internal planetary pressures.

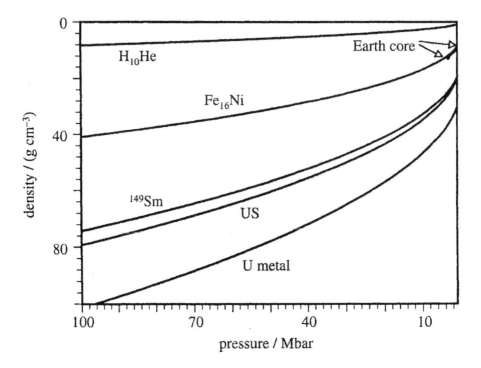

Figure 1. Theoretical pressure-density profiles of selected substances calculated using a Thomas-Fermi-Dirac approach (Salpeter & Zapolsky 1967). The applicability of the calculations relative to planetary interiors is demonstrated by the reasonable, although not perfect, agreement between the curve for Fe16Ni and the seismically-based estimate for the Earth's core (Dziewonski & Anderson 1981). Significantly, for the Thomas-Fermi-Dirac approach, errors are thought to decrease with increasing pressure (Stevenson & Salpeter 1976). This figure shows that uranium metal and uranium compounds, represented by uranium mono-sulphide (US), are more dense, at the pressures expected to prevail in planetary interiors, than any other substances including reactor poisons, as represented by the curve for 149Sm. Rather than uranium settling out directly from H-He, some U-containing complex of iron and other elements would be expected to settle out first, with uranium subsequently settling out from that complex.

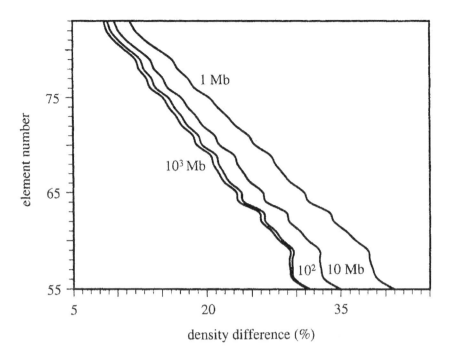

Figure 2 shows that, for elements with atomic numbers in the range 55Z83, i.e. the most heavy stable-elements, the percent differences in density relative to uranium are substantial, ranging from 9-41%; a high degree of separation would be anticipated. Witness, for example, the fact that the inner core of the Earth separated by the action of gravity from the fluid core even though the density difference is less than 5% (Dziewonski & Anderson 1981). Moreover, the greater the planet mass or protostar mass, the greater the gravitational acceleration and, consequently, the greater the degree of separation. In addition, it should be noted that, before the onset of nuclear reactions, convection will not take place.

In suggesting nuclear fission reactors as energy sources for the giant planets, I applied nuclear reactor theory (Fermi 1947) to demonstrate the feasibility that a concentration of uranium hydride might in the past have become critical, capable of sustaining a nuclear chain reaction. Continued, but interrupted, functioning as a breeder reactor was suggested based upon the behavior of the Oklo natural reactor. In calculating nuclear fission feasibility for the giant planets, I considered only slow (thermal) neutron fission; this paper reports the feasibility for fast neutron, non-hydrogenous, planetary and protostellar nuclear fission and discusses the implications of intermittent or interrupted reactor operation.

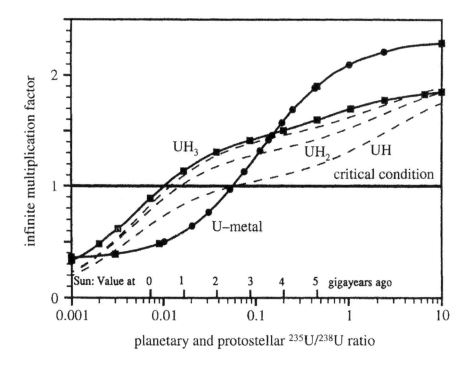

Figure 3. The infinite multiplication factor, k_∞, for fast and thermal neutron natural uranium reactors as a function of $^{235}U/^{238}U$ ratio. Corresponding times before present for terrestrial and presumed solar ratios are indicated as 'Sun value'. In a 'theoretically infinite' system, a nuclear chain reaction is possible when $k_\infty \geq 1$. Thermal neutron k_∞ curves for the indicated uranium hydride compounds are also shown. Solid k_∞ curves are based upon 27-group calculations which are the more sophisticated and accurate; 1-group results, indicated by dashed curves, are shown for comparison. At the time of formation of the Sun, between 4 and 5 billion years ago, the $^{235}U/^{238}U$ ratio was more than sufficient for an assemblage of a few kilograms or more of uranium to become supercritical and, gravitationally confined, to attain thermonuclear fusion ignition temperatures. Oak Ridge data points: •, Jordan & Turner (1992); □, D. F. Hollenbach (personal communication).

The fundamental criterion for maintaining a nuclear chain reaction is that on the average at least one neutron produced in a fission event causes another fission to occur. This criterion, referred to as criticality or critical condition, is described in nuclear reactor theory (Fermi 1947) by the unitary value of the neutron multiplication factor, k, where

$$k = k_\infty P \qquad (3.1)$$

P is a measure of the probability that neutrons will not be lost from the system and, being related to the geometry and mass of the reactor assembly, is always less than 1 except for an ideal, infinite assembly. For a system appropriate to planetary or protostar-scale reactors, P is approximately 1, so that

$$k = k_\infty \qquad (3.2)$$

The infinite multiplication factor, k_∞, is the ratio of the average number of neutrons produced in each generation to the average number of corresponding neutrons absorbed. As discussed by Herndon (1992), the expression for k_∞ from nuclear reactor theory, applicable to slow (thermal) neutron, hydrogenous, planetary-scale, nuclear fission reactors, is given by

$$k_\infty = \eta \varepsilon p f, \qquad (3.3)$$

where η is the average number of neutrons liberated for each neutron absorbed, ε is the fast fission factor, p is the resonance escape probability, and f is the thermal utilization factor. For fast neutron, non-hydrogenous, planetary-scale, planetary and protostellar nuclear fission reactors, equation (1.3) reduces to

$$k_\infty = \eta f, \qquad (3.4)$$

Methods for calculating the components of k_∞ are described in numerous textbooks (Foster & Wright 1973; Lamarsh 1983).

For a natural reactor, k_∞ depends upon the ratio $^{235}U/^{238}U$ which, through radioactive decay, changes over time and which may also change as a consequence of nuclear fission reactions. Figure 3 is a plot of k_∞, expressed as a function of the time before the present that natural, terrestrial uranium would have the indicated $^{235}U/^{238}U$ ratio, assuming no nuclear transmutation except through radioactive decay. Figure 3 presents the k_∞ curve for the fast neutron fission of uranium metal. Because the ratio of fission cross-section to capture cross-section is significantly greater at high neutron energies, the fast reactor k_∞ curve shown is approximately the same for various uranium compounds, including U metal, USi, US, UO_2, UC, and possibly others. The k_∞ curve for the slow neutron fission of uranium hydride, UH_3, shown in figure 3, serves as a useful reference.

As discussed previously (Herndon 1992), if a substantial quantity of UH_3 (at least several kilograms) were to have accumulated before about 500 million years ago, that mass would have begun to function as a thermal neutron reactor; continued operation to the present, however, would depend upon the nature of the fuel breeding reactions involved, e.g. $^{238}U(n, \gamma)$ $^{239}U(\beta\text{-})$ $^{239}Np(\beta\text{-})$ $^{239}Pu(\alpha)$ ^{235}U.

Similarly, *as* shown in figure 3, if a theoretically infinite quantity of uranium metal or, in fact, most compounds of uranium, were to have accumulated before about 2600 million years ago, that mass would have begun to function as a fast neutron reactor. Continued operation to the present, as in the thermal neutron case, would depend upon the nature of the fuel breeding reactions involved.

The importance of figure 3 is in showing that uranium, virtually irrespective of chemical state, would be capable of self-sustaining nuclear fission, if present in planetary cores and concentrated by gravity to the planets' centre before about 2600 million years ago. For protostars and the giant planets, the implication is that, even were the thermodynamic data on uranium hydrides to be incorrect, or if hydrogen was driven away from uranium by elevated temperatures, nuclear fission would nevertheless occur.

For the Sun, the requisite ratios, $^{235}U/^{238}U \geq 0.06$, have certainly existed during that star's lifetime, as inferred from terrestrial isotope ratios. Temporal specification based upon terrestrial isotope ratios is only relevant, within the framework of present knowledge, to the protostar that became the Sun. The k_∞ results presented in figure 3 are, however, generally valid for the indicated $^{235}U/^{238}U$ protostellar abundance ratios.

Approximately 2600 million years ago and earlier, the solar $^{235}U/^{238}U$ abundance ratio was sufficiently great for a 'theoretically infinite' uranium assemblage to become supercritical, as shown in figure 3 by the values of k_∞. Protostellar masses sufficient to gravitationally confine thermonuclear reactions, at least approximately 8% of the *mass* of the Sun, could likewise confine nuclear fission reactions and would permit the attainment of temperatures sufficiently high to ignite thermonuclear fusion reactions.

4. Implications of nuclear fission in the planetary sciences

Unlike previously envisioned planetary-scale energy sources that change gradually and in one direction through time, variable and interrupted energy output is possible from nuclear fission, as evidenced from investigations of the Oklo natural reactor (Maurette 1976). As known from nuclear reactor technology, various factors can shut down a nuclear reactor or can cause a nuclear reactor to run wild. One planetary nuclear reactor interruption mechanism envisioned relates to accumulations of reactor poisons effectively shutting down the geo-reactor for a period of time until the less dense reactor poisons diffuse from the region of the reactor sub-core (Herndon 1993). A similar mechanism may operate in the deep interiors of the giant planets and be the explanation of the differences in excessive luminosity referred to in § 2 above.

Changes occurring within the deep interior regions of the Earth are manifest as changes in the direction and/or intensity of the geomagnetic field and are evident over geologic time from palaeomagnetic investigations. Moreover, the consequences of such changes, although not yet understood, may affect surface phenomena, *as* suggested by apparent correlation of geomagnetic field reversals with species extinction (Hagiwara 1991; Kennett & Watkins 1970) and with major episodes of volcanism (Irvine 1989; Marzocchi 1990).

Likewise, changes may also be occurring within the giant planets; for example, during the past 120 years, significant variations have been noted in the appearance of turbulent features, particularly the Great Red Spot, in the atmosphere of Jupiter. In 1878 the Great Red Spot increased to a prominence not before recorded, but late in 1882 its prominence, darkness and general visibility began declining so steadily that by 1890 astronomers thought that the Great Red Spot was doomed to extinction (Peek 1958). Whether or not observed variations in Jupiter's turbulent features are due to changes in internal energy production is not known, but it is an interesting and important question.

Long-term monitoring indicates that gradual changes have occurred in the brightness of Uranus over the past thirty years (Lockwood *et al.* 1983). Continued monitoring is important to ascertain whether or not the observed brightening is solely a consequence of highly reflective polar regions and the 98° obliquity of that planet as suggested by some investigators (Conrath *et al.* 1991). Similar, long-term, more or less cyclic changes in the brightness of

Neptune have been observed for almost two decades (Lockwood & Thompson 1991). It is important to establish whether or not the atmospheric variations observed in the giant planets are related to changes in internal energy production.

5. Implications of nuclear fission in the astronomical sciences

The traditional concept of stellar ignition through temperatures developed by gravitational infall of matter and protostellar collapse dynamics assumes the inevitability of thermonuclear ignition, except for those protostars having masses less than approximately 8% of the mass of the Sun. Such very low mass objects, called brown dwarfs, are thought to approach minimum internal pressure limits for gravitational thermonuclear fusion confinement (Liebert & Probst 1987).

A considerable body of evidence has now been accumulated suggesting that the Universe contains at least ten times more non-luminous matter than luminous matter (Trimble 1987). The nature of dark matter is unknown and represents an outstanding problem in astrophysics.

Dark matter might be accounted for, at least in part, by the presence of dark stars, but not necessarily brown dwarfs, whose protostellar nuclear fission reactors failed to ignite thermonuclear fusion reactions. Possible reasons for such failure include a too low $^{235}U/^{238}U$ ratio, inadequate confinement pressure, and the absence of fissionable elements.

Observational evidence, primarily based on velocity dispersions and rotation curves, suggests that spiral galaxies have associated with them massive, spheroidal, dark matter components, thought to reside in their galactic halos (Rubin 1983). Interestingly, the luminous disc stars of spiral galaxies belong to the heavy-element-rich Population I; the luminous spheroidal stars of spiral galaxies belong to the heavy-element-poor Population II. In spiral galaxies, the dark matter components are thought to be associated in some manner with the spheroidal heavy-element-poor Population II stars (Bacall 1986; van der Kruit 1986). The association of dark matter with heavy-element-poor Population II stars is inferred to exist elsewhere, for example, surrounding elliptical galaxies (Jarvis & Freeman 1985; Levison & Richstone 1985). Because of the apparent association of dark matter with heavy-metal-poor Population II stars, I suggest the possibility that these dark matter components are composed of what might be called Population III stars, i.e. stars devoid of fissionable elements, and, consequently, unable to sustain the nuclear fission chain reactions necessary for the ignition of thermonuclear fusion reactions.

The possible existence of a second mechanism for stellar thermonuclear ignition precludes the notion that the sole cause of ignition was the heat generated by gravitational potential energy release. Observational confirmation of protostellar ignition is clearly important, not only as relates to the question of stellar ignition, but as relates to the possibility of stellar non-ignition and the nature of dark matter in the Universe.

References

Bacall, J. N. 1986 Star counts and galactic structure. *A. Rev. Astron. Astrophys.* 24, 577-611.

Baudin, G., Blain, C., Hagemann, R., Kremer, M., Lucas, M., Merlivat, L., Molina, R., Nief, G., Prost-Marechal, F., Regnaud, F. & Roth, E. 1972 *C. r. Acad. Sci., Paris* D275, 2291.

Bethe, H. A. 1939 Energy production in stars. *Phys. Rev.* 55, 434-456.

Bodu, R., Bouzigues, H., Morin, N. & Pfiffelmann, J. P. 1972 Sur l' existence anomalies isotopiques rencontrees dan l'uranium du Gabon. *C. r. Acad. Sci., Paris* D275, 1731-1736.

Conrath, B. J., Pearl, J. C., Appleby, J. F., Lindal, J. F., Orton, G. S. & Bezard, B. 1991 In *Uranus* (ed. J. T. Bergstralh, E. D. Miner & M.S. Mathews). Tucson: University of Arizona Press.

Dziewonski, A. M. & Anderson, D. A. 1981 Preliminary reference Earth model. *Phys. Earth Planet. Inter.* 25, 297-356.

Fermi, E. 1947 Elementary theory of the chain-reacting pile. *Science, Wash.* 105, 27-32.

Flügge, F. 1939 Kann der Energieinhalt der Atomkerne technisch nutzbar gemacht werden?*Naturwissenschaften* 27, 402.

Foster, A. R. & Wright, R. L. 1973 *Basic nuclear engineering.* Boston, Massachussetts: Allyn and Bacon.

Gamow, G. & Teller, E. 1938 The rate of selective thermonuclear reactions. *Phys. Rev.* 53, 608-609.

Hagiwara, Y. 1991 Geocatastrophy: mass extinction and geomagnetic reversal. *Chigaku Zasshi* 100, 1059.

Hahn, 0. & Strassmann, F. 1939 Uber den Nachweis und das Verhalten der bei der Best rahlung des Urans mittels Neutron entstehenden Erdalkalimetalle. *Naturwissenschaften* 27, 11.

Hayashi, C. & Nakano, T. 1965 Thermal and dynamical properties of a protostar and its contraction to the stage of quasi-static equilibrium. *Prog. theor. Physics* 35, 754-775.

Herndon, J. M. 1992 Nuclear fission reactors as energy sources for the giant outer planets. *Naturwissenschaften* 79, 7-14.

Herndon, J. M. 1993 Feasibility of a nuclear fission reactor at the center of the Earth as the energy source for the geomagnetic field. *J. Geomag. Geoelectr.* 45, 423-437.

Hubbard, W. B. 1990 In *The new solar system* (ed. J. K. Beatty & A. Chaikin), p. 131. Cambridge, Massachussetts: Sky.

Irvine, T. N. 1989 A global convection framework: concepts of symmetry, stratification, and system in the Earth's dynamic structure. *Econ. Geol.* 84, 2059-2114.

Jarvis, B. J. & Freeman, K. C. 1985 A dynamical model for galactic bulges. *Astrophys. J.* 295, 314-323.

Jordan, W. C. & Turner, J. C. 1992 Estimated critical conditions for UO_2F_2-H_2O systems in fully water-reflected spherical geometry. Oak Ridge National Laboratory TM-12292.

Kennett, J.P. & Watkins, N. D. 1970 Geomagnetic polarity change, volcanic maxima and faunal extinction in the south Pacific. *Nature, Lond* . 227, 930-934.

Khramov, A. N. 1982 *Paleomagnetology*. Heidelberg: Springer-Verlag.

Kuroda, P. K. 1956 On the nuclear physical stability of the uranium minerals. *J. chem. Phys.* 25, 781-782.

Lamarsh, J. R. 1983 *Introduction to nuclear engineering*. Reading: Addison-Wesley.

Larson, R. B. 1984 Gravitational torques and star formation. *Mon. Not. R. astr. Soc.* 206, 197-207.

Levee, R. D. 1953 A gravitationally contracting stellar model. *Astrophys. J.* 117, 200-210.

Levison, H. F. & Richstone, D. o. 1985 Scale-free models of highly flattened elliptical galaxies with massive halos. *Astrophys. J.* 295, 340-348.

Liebert, J. & Probst, R. G. 1987 Very low mass stars. *A. Rev. Astron. Astrophys.* 25, 473-519.

Lockwood, G. W., Lutz, B. L., Thompson, D. T. & Warnoc III, A. 1983 The albedo of Uranus. *Astrophys. J.* 266, 402-414.

Lockwood, G. W. & Thompson, D. T. 1991 Solar cycle relationship clouded by Neptune's sustained brightness maximum. *Nature, Lond.* 349, 593-594.

Marzocchi, W. 1990 Feasibility of a synchronized correlation between Hawaiian hot spot volcanism and geomagnetic polarity. *Geophys. Res. Lett.* 17, 1113-1116.

Maurette, M. 1976 Fossil nuclear reactors. *A. Rev. Nuc. Sci.* 26, 319-350.

Neuilly, M., Bussac, J., Frjacques, C., Nief, G., Vendryes, G. & Yvon, J. 1972 *C. r. Acad. Sci., Paris* D275, 1847.

Pearl, J. C., Conrath, B. J., Hanal, R. A. & Pirraglia, J. A. 1990 The albedo, effective temperature, and energy balance of Uranus, as determined from Voyager IRIS data. *Icarus* 84,12-28.

Peek, B. M. 1958 *The planet Jupiter*. London: Faber and Faber.

Rubin, V. C. 1983 The rotation of spiral galaxies. *Science, Wash.* 220, 1339-1344.

Salpeter, E. E. & Zapolsky, H. S. 1967 Theoretical high-pressure equations of state including correlation energy. *Phys. Rev.* 158, 876-886.

Schwartzschild, M. 1958 *Structure and evolution of the stars*. Princeton University Press.

Stevenson, D. J. & Salpeter, E. E. 1976 In *Jupiter*. (ed. T. Gehrcls), p. 85. Tucson: University of Arizona Press.

Trimble, V. 1987 Existence and nature of dark matter in the universe. *A. Rev. Astron. Astro phys.* 25, 425-472.

van der Kruit, P. C. 1986 Surface photometry of edge-on spiral galaxies. *Astron. Astrophys.* 157, 230-244.

Appendix F

Internal Heat Production in Hot Jupiter Exo-planets, Thermonuclear Ignition of Dark Galaxies, and the Basis for Galactic Luminous Star Distributions

J. Marvin Herndon
Transdyne Corporation

Published in

Current Science, Vol. 96, No. 11, 10 June 2009, pp. 1453-1456.
Download PDF http://www.nuclearplanet.com/1453.pdf

Astronomical observations of planets orbiting stars other than our Sun, will inevitably lead to a more precise understanding of our own Solar System and as well, perhaps, of the Universe as a whole. The discovery of so-called 'hot Jupiter' exo-planets, those with anomalously inflated size and low density relative to Jupiter, has evoked much discussion as to possible sources of internal heat production. But to date, no explanations have come forth that are generally applicable. For example, hot Jupiter exo-planets are found with insufficient eccentricity to be heated internally by tidal dissipation as originally suggested [1]. Other ideas, such as internal conversion of incident radiation into mechanical energy [2] and on-going tidal dissipation due to a non-zero planetary obliquity [3] also appear to lack general applicability. Charbonneau *et al.* [4] noted that two cases [HD 209458b and HAT-P-1b] suggest at least '...there is a source of internal heat that was overlooked by theoreticians'.

One purpose of the present note is to suggest a source of internal heat production for hot Jupiters exo-planets that indeed has been overlooked by theoreticians and which may be of general applicability. Another purpose is to suggest that the observation of hot Jupiter exo-planets may prove to be the first observational evidence of the correctness of my concept of the ignition of stellar thermonuclear fusion reactions by nuclear fission [5]. Yet another purpose is to discuss implications pertaining to the thermonuclear ignition of dark galaxies, and to suggest that the distributions of luminous stars in galaxies are reflections of the distributions of fissionable elements.

In the late 1960s, astronomers discovered that Jupiter radiates into space about twice as much energy as it receives from the sun. Later, Saturn and Neptune were also found to radiate prodigious quantities of internally generated energy. This excess energy production has been described as 'one of the most interesting revelations of modern planetary science' [6]. Stevenson [7], discussing Jupiter, stated, 'The implied energy source ... is apparently gravitational in origin, since all other proposed sources (for example, radio-activity, accretion, thermonuclear fusion) fall short by at least two orders of magnitude....' Similarly, more than a decade later, Hubbard [6] asserted, 'Therefore, by elimination, only one process could be responsible for the luminosities of Jupiter, Saturn, and Neptune. Energy is liberated when mass in a gravitationally bound object sinks closer to the centre of attraction ... potential energy becomes kinetic energy ....'

In 1990, when I first considered Jupiter's internal energy production, that explanation did not seem appropriate or relevant because about 98% of the mass of Jupiter is a mixture of hydrogen and helium, both of which are extremely good heat transport media. Moreover, the mass of Neptune is only about 5% that of Jupiter. Having knowledge of the fossil natural nuclear fission reactors that were discovered in 1972 at Oklo, Republic of Gabon, in Western Africa [8], I realized a different possibility and proposed the idea of planetary-scale nuclear fission reactors as energy sources for the giant planets [9]. At first I demonstrated the feasibility for thermal neutron reactors in part using Fermi's nuclear reactor theory, *i.e.*, the same calculations employed in the initial design of commercial nuclear reactors and used by Kuroda [10] to predict conditions for the natural reactors that were later discovered at Oklo. Subsequently, I extended the concept to include planetocentric fast neutron breeder reactors, which are applicable as well to non-hydrogenous planets, especially the nuclear georeactor as the energy source [5, 11] and the operant fluid [12, 13] for generating the earth's magnetic field.

There is a strong terrestrial evidence for the planetocentric nuclear reactor concept. In the 1960's geoscientists discovered occluded helium in oceanic basalts which, remarkably, possessed a higher $^3He/^4He$ ratio than air. At the time there was no known deep-earth mechanism that could account for the experimentally measured ^3He, so its origin was assumed to be a primordial ^3He component, trapped at the time of earth's formation, which was subsequently diluted with ^4He from radioactive decay. State-of-the-art numerical simulations of georeactor operation, conducted at Oak Ridge National Laboratory, USA, yielded fission-product helium, as shown in Figure 1, with isotopic compositions within the exact range of compositions typically observed in oceanic basalts [14, 15]. For additional information, see Rao [16].

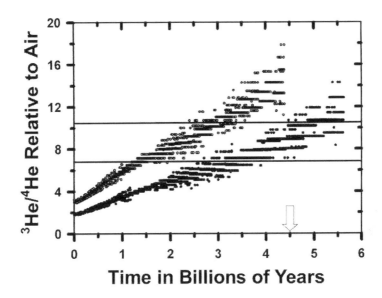

Figure 1. Fission product ratio of $^3He/^4He$, relative to that of air, from nuclear georeactor numerical calculations at 5 TW (upper) and 3 TW (lower) power levels, from [14]. The band comprising the 95% confidence level for measured values from mid-oceanic ridge basalts (MORB) is indicated by the solid lines. The age of the earth is marked by the arrow. Note the distribution of calculated values at 4.5 gigayears, the approximate age of the Earth. The increasing values are the consequence of uranium fuel burn-up. Iceland deep-source 'plume' basalts present values ranging as high as 37 [34].

At the pressures which exist near the centre of the Earth, density becomes a function almost exclusively of atomic number and atomic mass. Thus, heavy fission products, like krypton and xenon, are constrained to be trapped forever within the georeactor fission product sub-shell and will be unable to escape, never to be brought to earth's surface. Helium, on the other hand, can be expected to escape; the similarity in its isotopic composition with helium measured in oceanic basalts stands as evidence for the existence of the georeactor. In principle, another noble gas, neon, is sufficiently light so as to be able to escape from the earth's core, provided it can pass through the inner core. Neon, with a unique isotopic signature, is observed in deep-source basalts, such as those from Hawaii and Iceland [17]. A tantalizing possibility is that the observed neon is georeactor-produced. Regrettably, though, fission yield data on neon and its progenitor fission products are too scanty and imprecise to make such a determination.

There are two other potential possibilities for verifying the existence of the georeactor, but each presently lack sensitivity and resolution: seismic detection and anti-neutrino detection and discrimination [18].

At the beginning of the 20th century, understanding the nature of the energy source that powers the sun and other stars was one of the most important problems in physical science. Initially, gravitational potential energy release during protostellar contraction was considered, but calculations showed that the energy released would be insufficient to power a star for as long as life has existed on Earth. The discovery of radioactivity and the developments that followed led to the idea that thermonuclear fusion reactions power the sun and other stars.

Thermonuclear fusion reactions are called 'thermonuclear' because temperatures on the order of a million degrees Celsius are required. The principal energy released from the detonation of hydrogen bombs comes from thermonuclear fusion reactions. The high temperatures necessary to ignite H-bomb thermonuclear fusion reactions comes from their A-bomb nuclear fission triggers. Each hydrogen bomb is ignited by its own small nuclear fission A-bomb.

In 1938, when the idea of thermonuclear fusion reactions as the energy source for stars had been reasonably well developed [19], nuclear fission had not yet been discovered [20]. Astrophysicists assumed that the million-degree-temperatures necessary for stellar thermonuclear ignition would be produced by the in-fall of dust and gas during star formation and have continued to make that assumption to the present, although clearly there have been signs of potential trouble with the concept [21]. Proto-star heating by the in-fall of dust and gas is off-set by radiation from the surface which is a function of the fourth power of temperature. Generally, in numerical models of protostellar collapse, thermonuclear ignition temperatures, on the order of a million degrees Celsius, are not attained by the gravitational in-fall of matter without additional *ad hoc* assumptions, such as assuming an additional shockwave induced sudden flare-up [22] or result-optimizing the model-parameters, such as opacity and rate of in-fall [23].

After demonstrating the feasibility for planetocentric nuclear fission reactors, I suggested that thermonuclear fusion reactions in stars, as in hydrogen bombs, are ignited by self-sustaining, neutron induced, nuclear fission [5]. I now suggest the possibility that hot Jupiter exo-planets may derive much of their internal heat production from thermonuclear fusion reactions ignited by nuclear fission.

The discovery of hot Jupiter exo-planets has evoked much discussion as to possible sources of internal heat production, but to date no generally applicable astrophysical explanations have been presented.

One might expect planetocentric nuclear fission reactors to occur within exo-planets of other planetary systems that have a heavy element component, provided the initial actinide isotopic compositions are appropriate for criticality. And, indeed, planetocentric nuclear fission reactors may be a crucial component of hot Jupiter exo-planets. But it is unlikely that fission-generated heat alone would be sufficient to create the 'puffiness' that is apparently observed. For example, as calculated using Oak Ridge National Laboratory nuclear reactor numerical simulation software, a one Jupiter-mass exo-planet without any additional core enrichment of actinide elements could produce a constant fission-power output of $\sim 4 \times 10^{21}$ ergs/s for only $\sim 5 \times 10^8$ years. Even with that unrealistically brief interval, the fission-power output is orders of magnitude lower than the 10^{26} to 10^{29} ergs/s needed for the observed puffiness according to hot Jupiter model-calculations [1].

Unlike stars, hot Jupiter exo-planets are insufficiently massive to confine thermonuclear fusion reactions throughout a major portion of their gas envelopes. One might anticipate instead fusion reactions occurring at the interface of a central, internal substructure, presumably the exo-planetary core, which initially at least was heated to thermonuclear ignition temperatures predominantly by self-sustaining nuclear fission chain reactions. After the onset of fusion at that reactive interface, maintaining requisite thermonuclear-interface temperatures might be augmented to some extent by fusion-produced heat, which would as well expand the exo-planetary gas shell, thus decreasing the density of the exo-planet. Viewed in this context, hot Jupiter exo-planets appear to be stars in the process of ignition, at the cusp of being a star, but unable to fully ignite because their mass is almost, but not quite, sufficient for gravitational containment. Thus, observations of hot Jupiter exo-planets may stand as the first evidence for the correctness of my concept of stellar thermonuclear fusion ignition by nuclear fission chain reactions [5].

The idea that stars are ignited by nuclear fission triggers opens the possibility of stellar non-ignition, a concept which may have fundamental implications bearing on the nature of dark matter and, as suggested in the present note, on the thermonuclear ignition of dark galaxies, and on the distribution of luminous stars in galaxies Universe-wide. As I noted in 1994, the corollary to thermonuclear ignition is non-ignition, which might result from the absence of fissionable elements, and which would lead to dark stars [5].

Observational evidence, primarily based on velocity dispersions and rotation curves, suggests that spiral galaxies have associated with them massive, spheroidal, dark matter components, thought to reside in their galactic halos [24]. Interestingly, the luminous disc stars of spiral galaxies belong to the heavy-element-rich Population I; the luminous spheroidal stars of spiral galaxies belong to the heavy-element-poor Population Il. In spiral galaxies, the dark matter components are thought to be associated in some manner with the spheroidal heavy-element-poor Population II stars [25, 26]. The association of dark matter with heavy-element-poor Population II stars is inferred to exist elsewhere, for example, surrounding elliptical galaxies [27, 28]. Because of the apparent association of dark matter with heavy-metal-poor Population II stars, I have suggested the possibility that these dark matter components are composed of what might be called Population III stars, zero metallicity stars or stars at least devoid of fissionable elements, and, consequently, unable to sustain the nuclear fission chain reactions necessary for the ignition of thermonuclear fusion reactions.

Although dark matter is thought to be more than an order of magnitude more abundant than luminous matter in the Universe, there has yet to be an unambiguous identification of a wholly dark, galactic-scale structure. There is, however, increasing evidence that VIRGOHI 21, a mysterious hydrogen cloud in the Virgo Cluster, discovered by Davies et al. [29] may be a dark galaxy. Minchin et al. [30] suggested that possibility on the basis of its broad line width unaccompanied by any responsible visible massive object. Subsequently, Minchin et al.[31] find an indubitable interaction with NGC 4254 which they take as additional evidence of the massive nature of VIRGOHI 21. If indeed VIRGOHI 21 turns out to be composed of dark stars having approximately the mass of stars found in luminous galaxies, it would lend strong additional support to my concept of stellar thermonuclear ignition by nuclear fission [5].

The existence of a dark galaxy composed of non-brown-dwarf, solar-massive dark stars would certainly call into question the long-standing idea of gravitational collapse as the sole source of heat for inevitable stellar thermonuclear ignition, which after all has no laboratory support, unlike my idea of a nuclear fission trigger [5], which has been demonstrated experimentally with each H-bomb detonation.

For half a century, the concept that elements are synthesized within stars [32] has become widely accepted. In the so-called B²FH model, heavy elements are thought to be formed by rapid neutron capture, the R-process, at the supernova end of a star's lifetime; there may be another explanation.

The conditions and circumstances at galactic centres appear to harbor the necessary pressures for producing highly dense nuclear matter and the means to jet that nuclear matter out into the galaxy where, as suggested here, the jet seeds dark stars which it encounters with fissionable elements, turning dark stars into luminous stars. Galactic jets, either single or bi-directional, are observed originating from galactic centres, although little is currently known of their nature. Figure 2 is a Hubble Space Telescope image of a 10,000 light-year long galactic jet. One such jet was observed to have a length of 865,000 light years.

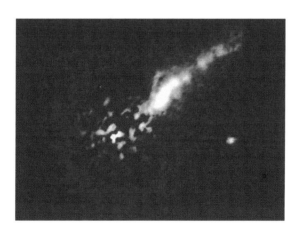

Figure 2. Hubble Space Telescope image of a 10,000 light year long galactic jet. Galaxy light was digitally removed for clarity. Galactic jets as long as 865,000 light years have been observed.

Figure 3. Hubble Space Telescope images of (a) anomalous galaxy NGC 4676, (b) anomalous galaxy, (c) spiral galaxy, M10 and (d) barred spiral galaxy, NGC 1300.

Consider a more-or-less spherical, gravitationally bound assemblage of dark (Population III) stars, a not-yet-ignited dark galaxy. Now, consider the galactic nucleus as it becomes massive and shoots its first jet of nuclear matter into the galaxy of dark stars, seeding and igniting those stars which it contacts. How might such a galaxy at that point appear? I suggest it would appear quite similar to NGC4676 (Figure 3 a) or to NGC10214 (Figure 3 b).

The arms of spiral galaxies, such as M101 (Figure 3 c), and the bars which often occur in disc galaxies [33], such as in NGC1300 (Figure 3 d), possess morphologies which I suggest occur as a consequence of galactic jetting of fissionable elements into the galaxy of dark stars, seeding the dark stars encountered with fissionable elements, thus making possible ignition of thermonuclear fusion reactions.

The structures of just about all luminous galaxies appear to have the jet-like luminous-star features, the imprints of the galactic jets which gave rise to their ignition, the imprints of the distribution of fissionable, heavy element seeds. Therein is the commonality connecting the diverse range of galactic observed structures and the causal relationship which appears to exist.

And what of the dark matter necessary for dynamical stability? The dark matter is the spherical halo of un-ignited, dark stars, located just where it must be to impart rotational stability to the galactic luminous structure [24].

Since the 1930s, astrophysics has been built upon the concept that thermonuclear reactions in stars are ignited automatically by heat generated by the collapse of dust and gas during star formation. Not only are there severe problems associated with that concept, because of extreme heat loss at high temperatures, but the observed jet-like distributions of luminous galactic stars are wholly inexplicable within that context. In stark contrast, the variety of morphological galactic forms, especially the prevalence of jet-like arms and bars can be understood in a logical and causally related way from my concept of heavy-elements being formed in galactic centres and jetted into space where they seed the dark stars they encounter with fissionable elements, which in turn ignite thermonuclear fusion reactions. From this perspective, the distribution of luminous stars in a galaxy, and consequently the type of galaxy, for example, barred or spiral, may simply be a reflection of the distribution of the fissionable elements jetted from the galactic centre.

References

1.   Bodenheimer, P., Lin, D. N. C. and Mardling, R. A., On the tidal inflation of short-period extrasolar planets. *Astrophys. J.*, 2001, **548**, 466-472.

2.   Showman, A. P. and Guillot, T., Atmospheric circulation and tides of "51 Pegasus-like" planets. *Astron. Astrophys.*, 2002, **385**, 166-180.

3.   Winn, J. N. and Holman, M. J., Obliquity tides on hot jupiters. *Astrophys. J.*, 2005, **625**, L159-L162.

4.   Charbonneau, D., et al., Precise radius estimates for the exoplanets WASP-1b and WASP-2b. , 2006, arXiv.org/astro-ph/0610589.

5.   Herndon, J. M., Planetary and protostellar nuclear fission: Implications for planetary change, stellar ignition and dark matter. *Proc. R. Soc. Lond.*, 1994. **A455**, 453-461.

6.   Hubbard, W. B., Interiors of the giant planets, in *The New Solar System*, A. Chaikin, Editor. 1990, Sky Publishing Corp.: Cambridge, MA. pp. 134-135.

7.   Stevenson, J. D., The outer planets and their satellites, in *The Origin of the Solar System*, S.F. Dermott, Editor. 1978, Wiley: New York. pp. 395-431.

8.   Bodu, R., et al., Sur l'existence anomalies isotopiques rencontrees dan l'uranium gu Gabon. *C. r. Acad. Sci., Paris*, 1972, **D275**, 1731-1736.

9.   Herndon, J. M., Nuclear fission reactors as energy sources for the giant outer planets. *Naturwissenschaften*, 1992, **79**, 7-14.

10.   Kuroda, P. K., On the nuclear physical stability of the uranium minerals. *J. Chem. Phys.*, 1956, **25**, 781-782.

11.   Herndon, J. M., Feasibility of a nuclear fission reactor at the centre of the Earth as the energy source for the geomagnetic field. *J. Geomag. Geoelectr.*, 1993, **45**, 423-437.

12.   Herndon, J. M., Nuclear georeactor generation of the Earth's geomagnetic field. *Curr. Sci.*, 2007, **93(11)**, 1485-1487.

13.   Herndon, J. M., *Maverick's Earth and Universe*. 2008, Vancouver: Trafford Publishing. ISBN 978-1-4251-4132-5.

14.   Herndon, J. M., Nuclear georeactor origin of oceanic basalt ^{3}He/^{4}He, evidence, and implications. *Proc. Nat. Acad. Sci. USA*, 2003, **100(6)**, 3047-3050.

15.   Hollenbach, D. F. and Herndon, J. M., Deep-earth reactor: nuclear fission, helium, and the geomagnetic field. *Proc. Nat. Acad. Sci. USA*, 2001, **98(20)**, 11085-11090.

16. Rao, K. R., Nuclear reactor at the core of the Earth! - A solution to the riddles of relative abundances of helium isotopes and geomagnetic field variability. *Curr. Sci.*, 2002, **82(2)**, 126-127.

17. Tieloff, M., et al., The nature of pristine noble gases in mantle plumes. Science, 2000, **288**, 1036-1038.

18. Raghavan, R. S., Detecting a nuclear fission reactor at the centre of the earth. 2002, arXiv:hep-ex/0208038.

19. Bethe, H. A., Energy production in stars. *Phys. Rev.*, 1939, **55(5)**, 434-456.

20. Hahn, O. and Strassmann, F., Uber den Nachweis und das Verhalten der bei der Bestrahlung des Urans mittels Neutronen entstehenden Erdalkalimetalle. *Die Naturwissenschaften*, 1939, **27**, 11-15.

21. Hayashi, C. and Nakano, T., Thermal and dynamic properties of a protostar and its contraction to the stage of quasi-static equilibrium. *Prog. theor. Physics*, 1965, **35**, 754-775.

22. Larson, R. B., Gravitational torques and star formation. *Mon. Not. R. astr. Soc.*, 1984, **206**, 197-207.

23. Stahler, S. W., et al., The early evolution of protostellar disks. *Astrophys. J.*, 1994, **431**, 341-358.

24. Rubin, V. C., The rotation of spiral galaxies. *Science*, 1983, **220**, 1339-1344.

25. Bacall, J. N., Star counts and galactic structure. *A. Rev. Astron. Astrophys.*, 1986, **24**, 577-611.

26. van der Kruit, P. C., Surface photometry of edge-on spiral galaxies. *Astron. Astrophys.*, 1986, **157**, 230-244.

27. Jarvis, B. J. and Freeman, K. C., A dynamical model for galactic bulges. *Astrophys. J.*, 1985, **295**, 314-323.

28. Levison, H. F. and Richstone, D.O., Scale-free models of highly flattened elliptical galaxies with massive halos. *Astrophys. J.*, 1985, **295**, 340-348.

29. Davies, J., et al., A multibeam HI survey of the Virgo Cluster - two isolated HI clouds? *Mon. Not. R. astr. Soc.*, 2004, **349(3)**, 922.

30. Minchin, R., et al., A dark hydrogen cloud in the Virgo structure. *Astrophys. J.*, 2005, **622**, L21-L24.

31. Minchin, R., et al., High resolution H i imaging of VIRGOHI 21 - a dark galaxy in the Virgo Cluster. *American Astronomical Society Meeting 207 #188.13*, 2005.

32. Burbidge, E. M., et al., Synthesis of the elements in stars. *Rev. Mod. Phys.*, 1957, **29(4)**, 547-650.

33. Gadotti, D. A., Barred galaxies: an observer's prospective. 2008, arXiv:0802.0495.

34. Hilton, D. R., et al., Extreme He-3/He-4 ratios in northwest Iceland: constraining the common component in mantle plumes. *Earth Planet. Sci. Lett.*, 1999, **173(1-2)**, 53-60.

Appendix G

J. Marvin Herndon's Scientific Advances in Geoscience, Planetary Science, and Astrophysics

Webpage http://www.NuclearPlanet.com/advances.html has links for downloading scientific articles and for connecting to various descriptions of the subject matter.

J. Marvin Herndon (b. 1944) is an American interdisciplinary scientist, who earned his BA degree in physics in 1970 from the University of California, San Diego and his Ph.D. degree in nuclear chemistry in 1974 from Texas A&M University. J. Marvin Herndon was a post-doctoral apprentice to Hans E. Suess and Harold C. Urey in geochemistry and cosmochemistry at the University of California, San Diego. He is the President and CEO of Transdyne Corporation in San Diego, California.

Profiled in 2003 in *Current Biography*, along with Chief Justice of the U. S. Supreme Court, William H. Rehnquist, White House chief of staff, Andrew H. Card, Jr., film director and screenwriter, Sofia Coppola and thirteen others, dubbed a "maverick geophysicist" by *The Washington Post* [*The Washington Post*, March 24, 2003, Page A06], and armed with a unique knowledge of the nature of science and the ways to make important discoveries, passed down through generations of master scientists, J. Marvin Herndon's professional life, as a technologist and as a scientist, has been a step-by-step logical progression of understanding and discovery, uncovering and posing corrections to deep-rooted mistakes in geophysics, in astrophysics, and in science management. His concept of a nuclear fission georeactor at Earth's center was the feature article, cover story of the August 2002 issue of *Discover* magazine. His work has been featured in news and magazine articles worldwide, for examples, from the *Sunday Times of London* to *Japanese Playboy* and from *Science & Vie* to *Newton*. His scientific insights, advances and discoveries follow:

Composition of Earth's inner core: On the basis of data discovered in the 1960's, J. Marvin Herndon deduced the composition of the inner core as being fully-crystallized nickel silicide, not partially-crystallized nickel-iron metal as proposed by Francis Birch in 1940. [Herndon, J. M. (1979) The

317

nickel silicide inner core of the Earth. ***Proceedings of the Royal Society of London***, A368, 495-500].

Enstatite chondritic composition of Earth's lower mantle and core: By fundamental ratios of mass, J. Marvin Herndon showed that the core and lower mantle of the Earth are chemically analogous to the two main components of the Abee enstatite chondrite. This provides evidence that the deep interior of the Earth is indeed like an enstatite chondrite meteorite and it means that one can estimate the abundances of the elements in the Earth's core and lower mantle from measured abundances in corresponding parts of the Abee meteorite [Herndon, J. M. (1980) The chemical composition of the interior shells of the Earth. ***Proceedings of the Royal Society of London***, A372, 149-154); Herndon, J. M. (2005) Scientific basis of knowledge on Earth's composition. ***Current Science***, **88**, 1034-1037; Herndon, J. M. (2011) Geodynamic basis of heat transport in the Earth. ***Current Science***, 101, 1440-1450].

Feasibility of a geocentric nuclear fission georeactor: With an understanding that the Earth's core contains uranium, J. Marvin Herndon applied Fermi's nuclear reactor theory to demonstrate the feasibility of a natural nuclear fission reactor at the center of the Earth, called the georeactor. Unlike other major, natural, Earth energy sources, which might change only gradually, the georeactor is capable of variable energy output including stopping (because of fission product accumulation) and re-starting again (as the light fission products migrate radially outward and uranium settles downward). Variable deep-Earth energy production may have important, not yet appreciated, implications on geomagnetic field variability and on planetary change [Herndon, J. M. (1993) Feasibility of a nuclear fission reactor at the center of the Earth as the energy source for the geomagnetic field. ***Journal of Geomagnetism and Geoelectricity***, **45**, 423-437; Herndon, J. M. (1994) Planetary and protostellar nuclear fission: Implications for planetary change, stellar ignition and dark matter. ***Proceedings of the Royal Society of London***, A455, 453-461].

Georeactor as the source of deep-Earth helium observed in oceanic basalt: Daniel F. Hollenbach and J. Marvin Herndon demonstrated, from numerical simulations made at Oak Ridge National Laboratory, that a deep-Earth nuclear fission reactor will produce both light-helium, ^{3}He, and heavy-helium, ^{4}He, precisely within the range of values observed from deep-source lavas. The helium found in oceanic lavas, first observed over three decades ago, is evidence that a planetary-scale, natural nuclear fission reactor operates at the center of the Earth [Hollenbach, D. F. and Herndon, J. M. (2001) Deep-earth reactor: nuclear fission, helium, and the geomagnetic field. *Proceedings of the National Academy of Sciences USA*, **98**, 11085-11090; Herndon, J. M. (2003) Nuclear georeactor origin of oceanic basalt 3He/4He, evidence, and implications. *Proceedings of the National Academy of Sciences USA*, **100**, 3047-3050].

Helium evidence of eventual georeactor demise: J. Marvin Herndon demonstrated, from numerical simulations made at Oak Ridge National Laboratory, that a deep-Earth nuclear fission reactor, the georeactor, will produce sufficient helium with precisely the range of ratios as observed from deep-source oceanic basalt lavas. Moreover, the ratio of 3He/4He increases over the lifetime of the georeactor. The high ratios observed in Icelandic and Hawaiian basalts suggest that the end of the georeactor lifetime is approaching, perhaps within the next billion years, perhaps much sooner; the time-frame is not yet known. Presumably, soon thereafter the geomagnetic field will begin its final collapse [Herndon, J. M. (2003) Nuclear georeactor origin of oceanic basalt 3He/4He, evidence, and implications. *Proceedings of the National Academy of Sciences USA*, **100**, 3047-3050].

Origin of the geomagnetic field: J. Marvin Herndon set forth a fundamentally new concept related to the generation of Earth's geomagnetic field. Previously, he had considered the nuclear reactor at the center of the Earth, the georeactor, only as the energy source for the dynamo mechanism which generates the geomagnetic field that is thought to arise from convective motions of an electrically conducting fluid in a rotating body. Since 1939, the operant fluid has been thought to be the Earth's fluid iron-alloy core. He suggested instead that the operant fluid may be contained within the georeactor as the fluid fission product and radioactive decay product sub-shell surrounding the actinide sub-core. He thus extended the georeactor concept by suggesting that the georeactor is both the energy source and the dynamo mechanism for generating the geomagnetic field. He pointed out the reasons why long-term, sustained convection appears more feasible within the georeactor sub-shell than within the Earth's fluid core

[Herndon, J. M. (2007) Nuclear georeactor generation of the earth's geomagnetic field. *Current Science*, **93**, 1485-1457].

Physical impossibility of Earth-core convection: Since 1939 convection has been assumed to occur in the Earth's fluid core. J. Marvin Herndon realized that, because of compression, the matter at the base of the fluid core is too dense to float to the top as a result of thermal expansion. Moreover, heat loss from the top of the core is inhibited as the core is wrapped in a thermally insulating blanket, the mantle. Thermal convection under those circumstances is physically impossible. Herndon also discovered that the Rayleigh Number, often used to justify convection, is inappropriate for both the core and the mantle, as the Rayleigh Number was derived for an incompressible fluid, a fluid of constant density except as modified by thermal expansion at the bottom. [Herndon, J. M. (2009) Uniqueness of Herndon's georeactor: Energy source and production mechanism for Earth's magnetic field. **arXiv**:0903.4622); Herndon, J. M. (2011) Geodynamic basis of heat transport in the Earth. *Current Science*, **101**, 1440-1450].

Georeactor review article: [Herndon, J. M. (2014) Terracentric nuclear fission georeactor: background, basis, feasibility, structure, evidence and geophysical implications. **Current Science**, 106(4), 528-541].

Earth formation by raining-out from within a giant gaseous protoplanet: From observations of matter, J. Marvin Herndon deduced the basis and reasons for understanding planetary formation in the Solar System mainly as the consequence of "raining out" from within giant gaseous protoplanets, leading to initial Earth formation as a gas giant Jupiter-like planet, a concept consistent with observations of close-to-star gas giant exoplanets in other planetary systems [Herndon, J. M. (2006) Solar System processes underlying planetary formation, geodynamics, and the georeactor. *Earth, Moon and Planets*, **99**, 53-99].

New Indivisible Geoscience Paradigm: From our planet's early formation as a Jupiter-like gas giant, J. Marvin Herndon has deduced: (1) Earth's internal composition and highly-reduced oxidation state; (2) Powerful new internal energy sources, protoplanetary energy of compression and georeactor nuclear fission energy; (3) Georeactor geomagnetic field generation; (4) Decompression-driven geodynamics that accounts for the myriad observations attributed to plate tectonics without requiring mantle convection, and; (5) Fold-mountain formation that does not necessarily involve plate collisions. [J. M. Herndon (2011) New Indivisible Geoscience Paradigm. **arXiv**:1107.2149; Herndon, J. M. (2011) Geodynamic basis of heat transport in the Earth. ***Current Science***, **101**, 1440-1450].

Whole-Earth Decompression Dynamics: J. Marvin Herndon set forth the principles of Whole-Earth Decompression Dynamics which unifies elements of plate tectonics theory and Earth expansion theory into a uniquely new self-consistent vision of global geodynamics, obviating the assumption of mantle convection [Herndon, J. M. (2005) Whole-Earth decompression dynamics. ***Current Science***, **89**, 1937-1941; Herndon, J. M. (2004) Protoplanetary Earth formation: further evidence and geophysical implications. **arXiv**:astro-ph/0408539]. Herndon described, as one of the consequences of Whole-Earth Decompression Dynamics, an unrecognized, different energy source for driving geodynamics, the stored energy of protoplanetary compression augmented by georeactor nuclear fission energy, and proposed a new mechanism for transporting heat within the Earth, called Mantle Decompression Therma Tsunami, which emplaces heat and pressure at the base of the crust, producing volcanoes and causing earthquakes [Herndon, J. M. (2006) Energy for geodynamics: Mantle decompression thermal tsunami. ***Current Science***, **90**, 1605-1606].

Origin of mountains and primary initiation of submarine canyons: From our planet's early formation as a Jupiter-like gas giant and resulting changes in surface curvature, J. Marvin Herndon has deduced a new concept for the formation of mountains characterized by folding that does not necessarily require continent collisions. [J. M. Herndon (2012) Origin of mountains and primary initiation of submarine canyons: the consequences of Earth's early formation as a Jupiter-like gas giant. ***Current Science***, **102**, 1370-1372].

Fictitious supercontinent cycles: Descriptions of phenomena, events, or processes made on the basis of problematic paradigms can be unreasonably complex (e.g. epicycles) or simply wrong (e.g. ultraviolet catastrophe). Supercontinent cycles, also called Wilson cycles, are, as J. Marvin Herndon asserts, artificial constructs, like epicycles. Herndon provides the basis for that assertion and describes published considerations from a fundamentally different, new, indivisible geoscience paradigm which obviate the necessity for assuming supercontinent cycles. [J. Marvin Herndon (2012) Fictitious

Supercontinent Cycles, **arXiv**: 1302.1425]. Submitted for posting on 30 January 2013; still "on hold" by arXiv administrators as of 1 November 2014.

Earth core precipitates at core-mantle-boundary: J. Marvin Herndon predicted low-density, high-temperature Earth core precipitates [CaS and MgS] floating atop the fluid core at the core-mantle boundary. These are an expected consequence of the enstatite-chondrite-alloy-like core, originally containing some calcium and some magnesium dissolved in the iron alloy and are responsible for the seismic "roughness" observed there [Herndon, J. M. (1993) Feasibility of a nuclear fission reactor at the center of the Earth as the energy source for the geomagnetic field. *Journal of Geomagnetism and Geoelectricity*, **45**, 423-437; Herndon, J. M. (1996) Sub-structure of the inner core of the earth. *Proceedings of the National Academy of Sciences USA*, **93**, 646-648; Herndon, J. M. (2005) Scientific basis of knowledge on Earth's composition. *Current Science*, **88**, 1034-1037; Herndon, J. M. (2011) Geodynamic basis of heat transport in the Earth. *Current Science*, **101**, 1440-1450].

Physical impossibility of Earth-mantle convection: Since the 1931 convection has been assumed to occur in the Earth's mantle. J. Marvin Herndon has discovered that, because of compression, the matter the base of the mantle is too dense to float to the top as a result of thermal expansion. The mantle is not devoid of viscous losses, as evidenced by earthquakes to depths of 660 km. Convection is physically impossible under those circumstances. Herndon also discovered that the Rayleigh Number, often used to justify convection, is inappropriate for the mantle, as the Rayleigh Number was derived for an incompressible fluid, a fluid of constant density. [Herndon, J. M. (2009) Uniqueness of Herndon's georeactor: Energy source and production mechanism for Earth's magnetic field. **arXiv**:0903.4622; Herndon, J. M. (2011) Geodynamic basis of heat transport in the Earth. *Current Science*, **101**, 1440-1450].

Enhanced prognosis for abiotic natural gas and petroleum: J. Marvin Herndon pointed out that the prognosis for vast natural resources from abiotic natural gas and petroleum resources, which depends critically on the nature and circumstances of Earth formation, has for decades been considered solely within the framework of the now-discredited, 'standard model of solar system formation'. Within the context of recent advances related to the formation of Earth, initially as a Jupiter-like gas giant, that prognosis is greatly enhanced for several reasons [Herndon, J. M. (2006) Enhanced prognosis for abiotic natural gas and petroleum resources. *Current Science*, **91**, 596-598].

Geological basis for petroleum and natural gas deposits: The geology of planet Earth according to Herndon's Whole-Earth Decompression Dynamics (WEDD) is primarily the consequence of two processes: **(1)** The progressive formation of surface cracks to increase surface area in response to decompression-increased planetary volume, and; **(2)** The progressive adjustment of surface curvature in response to decompression-increased planetary volume. Crustal fragmentation, called rifting, provides all of the crucial components for petroleum-deposit formation: basin, reservoir, source, and seal. Rifting causes the formation of deep basins, as presently occurring in the Afar triangle of Northeastern Africa. Augmented by georeactor heat channeled upwards from deep within the Earth, uplift from sub-surface swelling can sequester sea-flooded lands to form halite evaporite deposits, can lead to dome formation, and can make elevated land susceptible to erosion processes, thus providing sedimentary material for reservoir rock in-filling of basins. Moreover, crustal fragmentation potentially exposes deep basins to sources of abiotic mantle methane and, although still controversial, methane-derived hydrocarbons..[Herndon, J. M. (2010) Impact of recent discoveries on petroleum and natural gas exploration: emphasis on India. *Current Science*, **98**, 772-779].

Potentially significant source of error in magnetic paleolatitude determinations: Magnetic paleolatitude measurements, as J. Marvin Herndon has shown, may be subject significant errors if the magnetization was emplaced when the Earth's radius was smaller than at present. Moreover, paleo-pole calculations are meaningless for such a circumstance [J. M. Herndon (2011) Potentially significant source of error in magnetic paleolatitude determinations. *Current Science*, **101**, 277-278].

New indivisible planetary science paradigm: The origin of planets and matter of the Asteroid Belt, planets' internal compositions, and the reason why the inner planets have such different surface dynamics despite being compositionally similar. [J. M. Herndon (2013) New indivisible planetary science paradigm. *Current Science*, 105(4), 450-460].

Origin of planetary magnetic fields: Currently active internally generated magnetic fields have been detected in six planets (Mercury, Earth,

Jupiter, Saturn, Uranus, and Neptune) and in one satellite, Jupiter's moon Ganymede. Magnetized surface areas of Mars and the Moon indicate the former existence of internally generated magnetic fields in those bodies. Based upon the commonality of matter in the Solar System and common operating environments, J. Marvin Herndon suggested that planetary and satellite magnetic fields arise from the same georeactor-type assemblage which he suggested powers and provides the operant fluid for generating by dynamo action the Earth's magnetic field [Herndon, J. M. (2009) Nature of Planetary Matter and Magnetic Field Generation in the Solar System. *Current Science*, **96**, 1033-1039].

Elucidation of planetary formation processes: J. Marvin Herndon showed that only three processes, operant during the formation of the Solar System, are responsible for the diversity of matter in the Solar System and are directly responsible for planetary internal-structures, including planetocentric nuclear fission reactors, and for dynamical processes, including and especially, geodynamics [Herndon, J. M. (2006) Solar System processes underlying planetary formation, geodynamics, and the georeactor. *Earth, Moon and Planets*, **99**, 53-99].

Nuclear fission reactors as energy sources for gas giant planets: With knowledge of the ancient remains of the natural nuclear reactors discovered at Oklo in the Republic of Gabon in Africa in 1972, J. Marvin Herndon demonstrated the feasibility of planetocentric nuclear fission reactors as energy sources for the gas giant outer planets [Herndon, J. M. (1992) Nuclear fission reactors as energy sources for the giant outer planets. *Naturwissenschaften* 79, 7-14].

Origin of ordinary chondrite meteorites: J. Marvin Herndon discovered a fundamental relationship using published whole-rock chondrite molar Mg/Fe and Si/Fe ratios. This relationship admits the possibility that ordinary chondrite meteorites are derived from two components: one is a relatively undifferentiated, primitive component, oxidized like the CI or C1 carbonaceous chondrites; the other is a somewhat differentiated, planetary component, with oxidation state like the highly reduced enstatite chondrites. Such a picture would seem to explain for the ordinary chondrites, their major element compositions, their intermediate states of oxidation, and their ubiquitous deficiencies of refractory siderophile elements. Herndon suggested that the planetary component of ordinary chondrite formation consists of planet Mercury's missing complement of elements [Herndon, J. M. (2004) Ordinary chondrite formation from two components: Implied connection to planet Mercury. **arXiv**:astro-ph/0405298; Herndon, J. M. (2004) Mercury's protoplanetary mass. **arXiv**:astro-ph/0410009; Herndon, J. M. (2004) Total mass of ordinary chondrite material originally present in the Solar System. **arXiv**:astro-ph/0410242; Herndon, J. M. (2007) Discovery of fundamental mass ratio relationships of whole-rock chondritic major elements: Implications on ordinary chondrite formation and on planet Mercury's composition. *Current Science*, **93**, 394-399].

Hydrogen geysers explanation for recent volatile activity on Mercury: J. Marvin Herndon showed that Mercury's origin by condensing at high pressures and high temperatures would lead to incorporation of vast quantities of hydrogen in Mercury's core which would be released upon solidification producing the observed pits and reducing an iron compound to iron metal yielding the highly-reflective material observed. [J. M. Herndon (2012) Hydrogen Geysers: Explanation for observed evidence of geologically recent volatile-related activity on Mercury's surface. *Current Science*, **103**, 361-362].

Stellar ignition by nuclear fission Thermonuclear fusion reactions, thought to power the Sun and other stars, require temperatures on the order of one million degrees Celsius for ignition. Since the mid-1930s the assumption has been that such temperatures were obtained during the in-fall of dust and gas during star formation, but there are problems. In 1994, J. Marvin Herndon suggested that stellar fusion reactions may, in fact, be ignited by a central fission reactor in the same manner that a fusion bomb is triggered by a fission bomb. Rather than stars automatically igniting during formation, non-ignition may occur in absence of actinide elements, leading to the possibility of dark stars, dark matter, particularly surrounding luminous galaxies [Herndon, J. M. (1994) Planetary and protostellar nuclear fission: Implications for planetary change, stellar ignition and dark matter. *Proceedings of the Royal Society of London*, **A455**, 453-461].

Origin of diverse luminous galaxy structures: J. Marvin Herndon has suggested that the diverse luminous galaxy structures can be understood in a logical and causally related manner if heavy element synthesis is related to galactic jets which jet heavy nuclear matter from the galactic core into the galaxy of dark stars where it seeds the dark stars it encounters with fissionable elements turning dark stars into luminous stars [Herndon, J. M. (2006) Thermonuclear ignition of dark galaxies. arXiv:astro-ph/0604307; Herndon, J. M. (2008) *Maverick's Earth and Universe*, Vancouver: Trafford Press, ISBN: 978-1-4251-4132-5]; Herndon, J. M., (2009) New concept for internal heat production in hot Jupiter exo-planets, thermonuclear ignition of dark galaxies, and the basis for galactic luminous star distributions. *Current Science*, **96**, 1453-1456].

Planetary interfacial thermonuclear fusion: J. Marvin Herndon has suggested that hot Jupiter exoplanets, which have densities less than Jupiter, may derive much of their internal heat production from interfacial thermonuclear fusion ignited by nuclear fission [Herndon, J. M. (2006) New concept for internal heat production in hot Jupiter exoplanets. arXiv:astro-ph/0612603; Herndon, J. M. (2008) *Maverick's Earth and Universe*, Vancouver: Trafford Press, ISBN: 978-1-4251-4132-5; Herndon, J. M., (2009) New concept for internal heat production in hot Jupiter exo-planets, thermonuclear ignition of dark galaxies, and the basis for galactic luminous star distributions. *Current Science*, **96**, 1453-1456].

Evidence against planetary migration: J. Marvin Herndon has presented evidence against the astrophysical concept of planetary migration based upon evidence that Earth was at one time a close-to-Sun gas giant similar to Jupiter in mass and composition [Herndon, J. M. (2006) Evidence contrary to the existing exoplanet migration concept. **arXiv**:astro-ph/0612726].

arXiv:astro-ph/0612603; Herndon, J. M. (2008) *Maverick's Earth and Universe*, Vancouver: Trafford Press, ISBN: 978-1-4251-4132-5; Herndon, J. M., (2009) New concept for internal heat production in hot Jupiter exo-planets, thermonuclear ignition of dark galaxies, and the basis for galactic luminous star distributions. *Current Science*, **96**, 1453-1456].

Evidence against planetary migration: J. Marvin Herndon has presented evidence against the astrophysical concept of planetary migration based upon evidence that Earth was at one time a close-to-Sun gas giant similar to Jupiter in mass and composition [Herndon, J. M. (2006) Evidence contrary to the existing exoplanet migration concept. **arXiv**:astro-ph/0612726].

Appendix G

Made in the USA
San Bernardino, CA
24 March 2016